CHEMISTRY AROUND YOU

Experiments and Projects with Everyday Products

Salvatore Tocci

East Hampton High School
East Hampton, New York

ARCO PUBLISHING, INC.
NEW YORK

For Nancy, Chris, Kerri, and Belle

Illustrations by Mary Jo Tocci

Published by Arco Publishing, Inc.
215 Park Avenue South, New York, N.Y. 10003

**Library of Congress Cataloging in Publication
Data**
Tocci, Salvatore.
  Chemistry around you.

  1. Chemistry—Experiments. I. Title.
QD38.T63 1985   540'.724   85-3881
ISBN 0-668-06026-3 (Cloth Edition)
ISBN 0-668-06033-6 (Paper Edition)

Printed in the United States of America

The Publisher wishes to acknowledge the assistance of Dave
Forman in the preparation of this book.

10   9   8   7   6   5   4   3   2   1

Contents

What in the World Isn't Chemistry?

Although you may not realize it, chemistry is an important part of your daily life. Chemistry is not limited to people working in laboratories; you deal with chemical principles, concepts, and reactions whenever you cook, take aspirin tablets, or wash clothes. Your home is, in fact, a chemical laboratory, providing you with an opportunity to observe, record, explain, and discover the world of chemistry as it operates in your kitchen, bathroom, laundry room, garage, and backyard. Many of the projects in this book are designed to make you an educated consumer who will buy products not on the basis of advertising gimmicks or manufacturers' claims but on the knowledge gained from your experimental results. With the information in this book, you'll know how to determine which orange juice provides the most vitamin C, which antacid is best for relieving acid indigestion, which detergent is most effective in cleaning clothes, which commercial fertilizer provides the most plant nutrients, and which consumer products are the best buys for many other household needs.

All these discoveries can be made with apparatus and chemicals commonly found in a school laboratory. In a few cases, an investigation can be performed more rapidly or an analysis made more thoroughly if some specialized equipment is available. For example, a meter to record the levels of acid in a solution will provide faster results than those obtained with testing papers. However, the same conclusions and discoveries are still possible without the meter. Consequently, the usual complement of beakers, graduated cylinders, flasks, Bunsen burners, and balances, coupled with a few chemicals, will be sufficient to perform the investigative exercises in this book.

Simple, But Effective Procedures

Procedures for weighing materials and performing other simple tasks are not explained. Whenever a technique or procedure not normally encountered outside a chemistry laboratory is used, the necessary information on how to proceed is given. Moreover, illustrations and diagrams are included to demonstrate many of the more involved procedures. One simple method requiring an explanation is the preparation of molar or normal solutions. In several chapters, you will discover projects involving 3M (molar) or 1N (normal) solutions. If these are not commercially available in solution form, prepare a 1M or 1N solution by dissolving an amount equal to the formula weight of the chemical substance (as given on the label) in distilled water to make 1 liter of solution. For a 2M or 2N solution, double the amount per liter. For example, a 1N sodium hydroxide solution is prepared by dissolving 40 gm, which is 1 formula weight, in enough distilled water to make 1 liter; a 2N by dissolving 80 gm, or 2 formula weights, to make 1 liter; a 3N, 120 gm, or 3 formula weights, to make 1 liter; and so on.

No matter how simple or complex the procedure, always observe laboratory safety procedures! Always wear safety goggles, avoid touching chemicals, never place flammable liquids near a flame, and know what you are doing at all times. The safest bet is to follow the instructions in this book for each project; all the necessary precautions are indicated. If in doubt, check with your teacher or a chemistry reference text.

Don't Stop Here

This book is meant as an introduction to the world of consumer chemistry. For those wishing to explore a certain concept or principle in more detail or to branch out to related areas, each chapter concludes with topics for further investigation. These ideas are suggestions, indicating additional problems where a knowledge of chemistry is required to search for the answers. Each chapter supplies the basic information; you must provide the initiative to see if any solutions can be found to the questions and problems posed by these topics for further investigation. For those seeking a greater challenge, some topics are indicated with asterisks (***). These represent problems of a greater magnitude, requiring a more sophisticated approach and more detailed knowledge of chemical concepts. In fact, these topics can serve as the basis of a valid scientific experiment, representing an original approach to the problem. Perhaps your Nobel prize in chemistry can be found in one of these topics!

Calculate, Conclude, and Compute

Many of the projects in this book require calculations before arriving at any conclusions. The mathematics is simple and carefully explained. However, if a microcomputer is available, you can enter the programs listed throughout this book to help you conduct the analysis. All the programs have been designed to run without any modification on the Apple II, Radio Shack TRS-80, and Commodore PET systems. Consequently, features unique to each microcomputer could not be utilized. If you have the knowledge, modify the programs so that they run more efficiently on your system. In addition, you may want to clear the computer screen at various points in the running of a program. But, if you're new to computers, don't worry. The programs are optional and not required for deriving the maximum benefit from the consumer projects. If you would like to get started, simply ask a teacher or friend for help in getting these listings translated into a running program. Once you get started, you'll discover how easy this process is. In fact, you may want to modify these listings as you proceed through the book to reflect additional computing ideas as they are introduced. Be aware that when running these programs, you should enter only numerical values and not units, that is 25 and not 25 gm, 3 and not 3 ml, etc. Once the numbers have been entered, the computer will do the rest!

CHEMISTRY IN THE KITCHEN

1 Cooks and Chemistry

Anyone who prepares meals in the kitchen is not only a cook but also a chemist! Whether the meal consists of hot dogs, spaghetti, or pork chops, foods are basically chemical substances known as *compounds*. Undoubtedly, you realize that cooking food usually makes it better tasting, easier to digest, and healthier. But did you know that cooking also causes chemical changes in the food, leading to the formation of new compounds? Whenever a new compound is made, a chemical reaction has occurred. Every time food is cooked, chemical reactions take place as the compounds in the food are changed into different substances. You can easily recognize that a chemical reaction has occurred by noting the changes in the appearance, feel, or taste of foods after they've been cooked.

While working in the kitchen, a cook may deal with many different chemical substances. These may include the spices, flavorings, and powders that are used in preparing meals. Actually, a good chef can be compared to a chemist doing research work in a laboratory—both may be experimenting with ways of making new substances, as pictured in Figure 1–1. Even preparing a simple meal, however, involves some chemistry. In fact, the most common chemical compound used in the kitchen is one of the most unusual substances known to chemists. This extraordinary chemical substance is water!

1

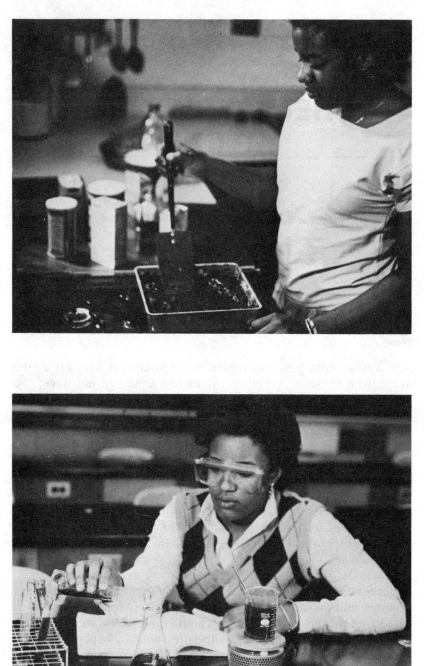

Figure 1–1. Whether cooking in a home economics class or conducting an experiment in a science laboratory, a student can be considered a chemist.

Water—An Unusual Substance

Since water is very common, we often take it for granted and fail to recognize its importance. Yet by weight, water makes up nearly 70 percent of the human body, where it plays an important role in the chemical reactions needed to keep us healthy. Although you could live for two months without food, you would die within a week without water. Even though water is both common and important, it has sometimes been called a "chemical freak" because it has several unusual properties.

For example, water is the only chemical substance that can exist simultaneously in three different forms or states under normal climatic conditions. Depending upon the temperature, water can be a solid (ice), liquid (rain), or gas (vapor produced by the heat from the sun). By the way, the temperatures required to change water from one form to another were important in developing the thermometer that you will be using in many of the laboratory exercises found in this book. In the eighteenth century, Anders Celsius, a Swedish scientist, placed an unmarked thermometer in melting ice and then in boiling water. After determining both the freezing and boiling points of water, he calibrated the thermometer. This was the first Celsius thermometer, which became the standard for use in all scientific laboratories throughout the world.

You can determine the temperatures required to change water from one form to another in the following experiments. You will be measuring two properties of water—its freezing point (the temperature at which water changes from a liquid into a solid) and its boiling point (the temperature at which it changes from a liquid into a gas). In addition, you will gain a working knowledge of the Celsius scale, which is the basis of all temperature measurements that you will record in performing your laboratory exercises.

Freezing Point Determination

1. Mix equal parts of crushed ice and coarse or rock salt in a styrofoam cup. Place a test tube containing some water in the cup. Make sure the level of the ice–salt mixture is above the level of the water in the test tube (see Figure 1–2).
2. Insert a Celsius thermometer into the test tube. Occasionally stir the ice–salt mixture by gently swirling the test tube in the cup.
3. Observe the water in the test tube to determine when ice begins to form. Record the temperature at which there is still some

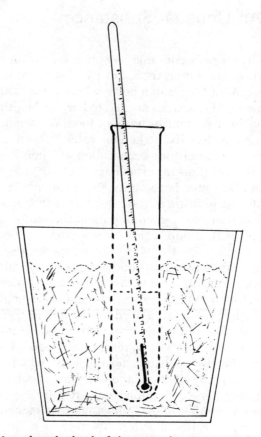

Figure 1–2. Note that the level of the ice–salt mixture in the cup must be higher than that of the water in the test tube.

liquid water left with the ice. The temperature at which ice forms is the freezing point of water.

4. Remove the thermometer from the test tube and place it in the ice–salt mixture. Record the lowest temperature. How does this temperature compare with the freezing point of water?

The Colder, the Better

You may have been surprised when comparing the temperature of the ice–salt mixture with the freezing point of water. Obviously, the addition of salt affected the temperature of the ice. Enjoy one benefit derived from this effect by using the following recipe to make a tasty treat!

1. Place an equal mixture of salt and crushed ice in a container large enough to hold a drinking cup.
2. In a mixing bowl, beat a whole egg, a tablespoon of sugar, and a dash of vanilla flavoring in a cup of whole milk.
3. Place a drinking cup almost full of the egg–milk mixture into the ice–salt until it freezes. Make sure the level of the ice–salt mixture is higher than that of the liquid in the cup. If the egg–milk mixture does not completely freeze by the end of class, remove the cup from the container and place it in a freezer until the next class meeting.
4. When the mixture has frozen, taste the product you have just prepared. Can you identify what you have made? Can you think of ways to change this recipe to create some interesting flavors?
5. Explain why the salt was added to the ice. Based on your observations in following this recipe, can you explain why rock salt is sometimes added to icy roads in winter? What effect would salt have upon the temperature at which the water would freeze?

Boiling Point Determination

1. Add several boiling chips to a flask approximately half-full of water.
2. Moisten the lower half of a Celsius thermometer with a lubricant such as glycerin. While holding the thermometer near the bulb, *gently* twist it into a one-hole rubber or cork stopper. DO NOT FORCE THE THERMOMETER INTO THE HOLE, SINCE IT MAY BREAK AND CAUSE A DEEP CUT. Clamp the stopper and thermometer to a ring stand and position the bulb of the thermometer so that it is completely submerged in the water. Place the stopper so that it does not seal or block the opening of the flask (see Figure 1–3).
3. Heat the water to boiling; the chips allow the heat to be evenly distributed, preventing the flask from shattering. Record the temperature when the water comes to a full boil. The temperature where the liquid changes into a gas is the boiling point.

 ## Celsius–Fahrenheit Conversions

As you noticed in the preceding experiment, the laboratory exercises in this book will involve the metric system of measurements. If you are not familiar with working with temperatures in degrees Celsius, you may want to type the following program into your computer. This program will demonstrate the relationship between the Fahrenheit and Celsius temperature scales. If you

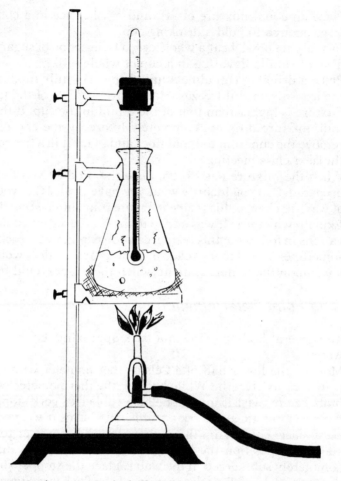

Figure 1–3. In determining the boiling point of water, make sure the bulb of the thermometer is immersed in the water. To avoid building up gas pressure, position the stopper so it does not block the flask opening.

wish to clear the screen at any point, you must use system specific commands, that is HOME for the Apple II and CLS for the TRS-80. Check with your teacher or a manual if you are not using an Apple II or TRS-80 microcomputer for the appropriate screen-clearing commands for your system.

```
10 PRINT "TO CONVERT A TEMPERATURE,"
20 PRINT "CHOOSE ONE OF THE FOLLOWING:":PRINT
30 PRINT "1) DEGREES FAHRENHEIT TO CELSIUS"
40 PRINT "2) DEGREES CELSIUS TO FAHRENHEIT"
50 PRINT "3) QUIT":PRINT
60 PRINT "WHICH NUMBER";:INPUT CH
70 IF CH = 2 THEN 160
80 IF CH = 3 THEN END
90 PRINT "WHAT IS THE TEMPERATURE"
100 PRINT "IN DEGREES FAHRENHEIT";:INPUT FA
110 CE = 5*(FA − 32)/9
120 PRINT
130 PRINT FA;" DEGREES FAHRENHEIT IS"
140 PRINT CE;" DEGREES CELSIUS."
150 PRINT:GOTO 10
160 PRINT "WHAT IS THE TEMPERATURE"
170 PRINT "IN DEGREES CELSIUS";:INPUT CE
180 FA = (9*CE)/5 + 32
190 PRINT
200 PRINT CE;" DEGREES CELSIUS IS"
210 PRINT FA;" DEGREES FAHRENHEIT."
220 PRINT:GOTO 10
```

Turning Water into Ice

In determining the freezing point of water, you were dealing with another of its unusual properties—its tendency to expand when it freezes. As the temperature drops, most liquids begin to contract as the particles that make up the liquid move closer together. When the liquid has turned completely into a solid, these particles have been packed into a smaller area or volume. Water, on the other hand, expands as it cools, causing the ice that is formed to take up a larger volume. Since the same number of particles are occupying a greater volume, the *density* of ice is less than that of water.

Density is defined as the mass (usually expressed as the weight) that is present in a defined volume. Because water expands as it freezes, the ice occupies a greater volume. Since the same number of water particles are in the ice as were present in the water, the mass (weight) is the same. However, since the volume they occupy is greater, the density of ice is less than that of water. With most chemical compounds, a liquid becomes denser as it turns into a solid. Therefore, a solid usually sinks in its own liquid. However, since ice is less dense than water, it floats. This is fortunate for both the plant life and the animal life living near the bottom of ponds and lakes. If ice were denser, these bodies of water would start to freeze at the bottom. Even in milder weather, the temperature of the deeper water might not get warm enough to melt the ice. This could be fatal for the organisms living near the bottom of ponds and lakes.

The fact that ice floats is proof that it is less dense than water. Since water expands approximately 10 percent of its volume when it freezes, only 10 percent of an iceberg is visible above the surface of the water. Nearly 90 percent of the iceberg is hidden beneath the surface of the ocean, presenting a serious danger to any ship that approaches. Can you think of an experiment you can perform in the laboratory to prove that water becomes less dense as it freezes? One way to demonstrate that ice is less dense than water is to examine what happens to the volume of water as it freezes. Perform the following experiment to discover what occurs as water turns into ice.

Water Versus Ice—Who's the Heavyweight?

1. Completely fill a small screw-top jar to the brim with water.
2. Securely tighten the top, and seal the jar inside a plastic container.
3. Put the container in a freezer for at least 24 hours and examine the jar the next day.
4. After comparing your results with those pictured in Figure 1–4, explain your observations. What happened to the volume, weight, and density of the water after it turned into ice?

Water Can Take the Heat

If you live near an ocean or any large body of water, you benefit from another unusual property of water—its tendency to absorb heat quickly without changing its own temperature very much. The temperatures of most liquids rise fairly rapidly as they are heated. Water, however, absorbs the heat rather rapidly so that its temperature does not increase as quickly. Consequently, on hot summer days the large amount of water in oceans or lakes can

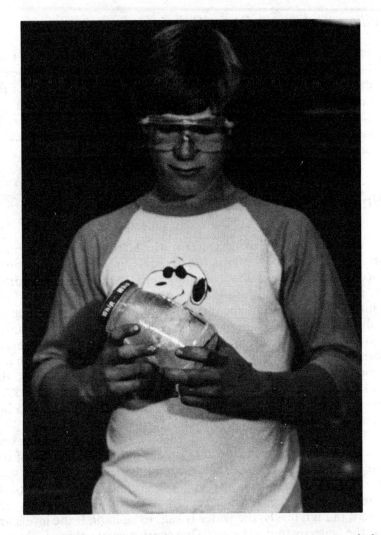

Figure 1–4. Can you explain why the jar this student is holding cracked?

absorb the heat from the sun, making the air temperatures cooler near the shore. Later in the year, this heat is slowly released from the water, causing the temperatures to be slightly warmer in coastal regions. You can verify this unusual ability of water to absorb heat rather rapidly by conducting the following exercise. By boiling water in a paper cup, not only will you prove that water absorbs heat more quickly than paper but you will also perform some chemical "magic"!

Boiling Water in a Paper Cup

1. Place an unwaxed paper cup on a ring stand. The cup must not be coated with wax because the wax will melt from the heat.
2. Completely fill the cup to the brim with water; any part of the cup that does not contain water will char.
3. *Gently* heat the paper cup by passing a Bunsen burner over the entire surface of the cup and continue heating until the water boils.

What's in Water Besides Water?

Whether you are using water in the kitchen or in the laboratory, most likely the water is not 100 percent pure. Water from a faucet contains many dissolved materials, including salts, minerals, gases, and natural pollutants such as decayed animals and plants. In addition, chemicals are sometimes added to water for health reasons. For example, you may live in an area where small amounts of fluoride are placed in drinking water to help prevent tooth decay. Chlorine gas is also usually added to water before it is pumped from reservoirs into homes. Because it is poisonous, the gas kills any harmful bacteria present in the drinking water. The amount of chlorine added is too small to cause any harm to humans.

If your drinking water comes from a well, you may have experienced an unpleasant taste; this is caused by a high level of the salts and minerals that can seep into the underground water supply. As the water passes through the ground, it picks up these chemical substances. Whenever these salts and minerals are present in high levels, the water is said to be *hard*. If the levels are low, the water is said to be *soft*. One characteristic of hard water is its inability to form suds when soaps or detergents are added. Hard water may also clog pipes and drains because the minerals and salts gradually form solid deposits that can block the flow of water.

Before reaching homes in a community that receives its water from reservoirs, the water is treated to reduce its hardness. The water is passed through filters containing chemicals that trap the salts and minerals, making the water soft. In homes where the well water is hard, commercially available softeners can be installed to soften the water. Since they contain chemicals to remove minerals from the water, these softeners are called demineralizers. These softeners may also contain activated charcoal to remove some of the dissolved gases and colored chemicals that can give water an unpleasant appearance, odor, or taste.

In the next experiment, you can examine the chemical difference between hard water and soft water. You will first prepare some hard water, which you will then treat with a demineralizer. Finally, you will check to see how successful you were in turning hard water into soft water.

Making Your Own Water Purifier

1. To prepare hard water, begin by placing 1 liter of water in a beaker. Add food coloring dye drop by drop, while stirring, until a *pale* color is visible. Add household ammonia drop by drop, while stirring, until the water has a *faint* odor of ammonia. Dissolve 5 gm of calcium chloride and 5 gm of magnesium chloride in the water. Record the appearance and odor of the hard water sample that you have just prepared.
2. Empty the contents of a small water demineralizer (those used for steam irons are suitable) into a beaker.
3. Add an equal amount of powdered activated charcoal to the contents of the demineralizer and mix well. If powdered charcoal is not available, you can crush some charcoal that is sold for aquarium filters.
4. Set up a filter apparatus (Figure 1–5) by placing the demineralizer–charcoal mixture into a funnel fitted with a piece of filter paper. Filter approximately half of the water sample through this apparatus and collect the liquid that passes through the filter in a flask. If the liquid is still colored or has an odor, keep passing the water back through the filter until the water is both colorless and odorless.
5. To test the softness of the filtered water, begin by placing three to four drops of liquid soap in each of two test tubes. Half fill one test tube with the original water sample and the other tube with the filtered water. Place a thumb on the top of each test tube and shake both of them vigorously for one minute. Measure the height of the suds formed in each of the test tubes.
6. Explain your results. If the height of the suds is the same in both tubes, refilter the water until a noticeable difference is observed.
7. You can check the relative hardness of samples brought from home by comparing the amount of suds produced with those obtained in this experiment. Explain why demineralizers are sold for steam irons.

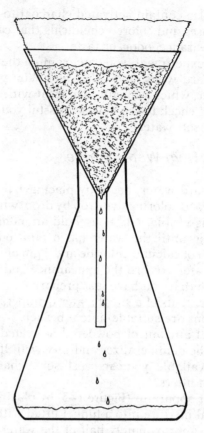

Figure 1–5. When filtering the hard water sample, be sure the level of the demineralizer–charcoal mixture is below that of the funnel.

Chemists are Picky People

As the water passed through the demineralizer–charcoal mixture, many of the dissolved chemicals were removed, producing a softer water that forms suds more easily. Water from which the salts and minerals have been removed is said to be *deionized*. However, deionized water is unsuitable for many of the experiments you will be conducting since the chemicals still present can interfere with the chemical reaction under study. In such situations, you must use *distilled* water, 100 percent pure water containing no minerals, salts, or any other chemicals. You can obtain distilled water at a pharmacy or you can prepare your own, using special laboratory apparatus. If you plan to use the distillation process, check a chemistry book to see how a Liebig condenser is set up to obtain distilled water by condensing the vapor produced from boiling tap water.

Topics for Further Investigation

The property of water to absorb heat rapidly is useful for solar heating systems. The heat energy from the sun is trapped by solar panels, where it is transferred to water. Examine the various ways solar energy systems operate. Could you design, build, and operate a solar energy system for some small-scale operation in either your home or school?

Most public water systems use various chemicals to kill small organisms that can cause disease in humans. However, not all the bacteria and viruses are killed before the water reaches the community. Can you design an experiment to determine the different kinds and numbers of bacteria present in tap water? You may wish to ask a biology teacher for information about isolating and growing bacteria.

In many parts of the world, the availability of water suitable for drinking is a problem. This is most apparent in arid regions such as deserts. What solutions can you propose to remedy this situation?

Nations have debated the ownership of the oceans' resources, from fish to mineral deposits. If you were a member of the World Court, what would be your decision on this issue? What about those coastal countries that do not have the technology to tap these resources?

Desalination of sea water has been suggested as a means of providing drinking water. Trace the history of this process and examine whether it represents a feasible approach to obtain fresh water.

Trace the route that water has taken to reach your home. Was it treated in any way before you used or drank it?

A critical water shortage has been declared in your community. As the director of the public water supply, what are your orders as to how water will be used during this crisis? What long-term solutions will you propose?

*** Water plays an important role in photosynthesis, serving as a source of oxygen required by most organisms, both plants and animals. Yet scientists know little about the chemical process by which photosynthetic plants use light energy to release the oxygen from water. Investigate one or more of the biochemical reactions in this process.

*** Aerobic respiration involving the use of oxygen depends upon a series of biochemical steps. The end result is the conversion of oxygen into water. During this process, most of the energy required by the organism is produced and stored in the form of adenosine triphosphate, abbreviated ATP. Plan a research project to examine how the production of ATP is linked to the conversion of oxygen into water during aerobic respiration.

2 Water—A Solution for Anyone's Problems

Remember the morning when you were so tired that you had trouble waking up to get ready for school? To help get going, you may have staggered into the kitchen to prepare some instant coffee. After putting a teaspoon of coffee granules in the cup, you added hot water and probably some sugar and milk. As you stirred the coffee, both the granules and the sugar disappeared. However, you knew by the taste that the coffee and sugar were still there. Obviously, stirring the coffee caused these particles to dissolve in the water until they were too small to see. Whenever a substance completely dissolves in a liquid so that the particles become invisible, a *solution* has been prepared.

Water can dissolve many different substances to form solutions. In the kitchen, you may have used water to dissolve not only coffee and sugar but also salt, powdered soft drinks, or packaged soups. The substance that does the dissolving, in this case water, is called the *solvent*. The substance that is dissolved, the sugar or salt, is known as the *solute*. Since water can dissolve so many different solutes to form solutions, chemists refer to water as a *universal solvent*. Although water is the most common solvent, other products often found in a kitchen are used to dissolve substances. For example, ammonia is useful for dissolving the grease that may splatter on stove and counter tops from cooking. Another solvent you might find in a kitchen is in the aerosol spray can used to dissolve the adhesive stuck to the skin from a bandage that has

15

been left on for a long time. With either the ammonia or the spray, a solvent is being used to dissolve a solute to make a solution.

A solution, however, may consist of several solutes dissolved in more than one solvent. This is the type of solution made in preparing instant coffee. Both the coffee granules and the sugar are dissolved in hot water and milk. In fact, milk itself may be considered a solution, consisting of a solvent (water) containing several solutes. Whenever the solutes are uniformly spread throughout a solvent, as they are in milk, the solution is said to be *homogeneous*.

If you examine the label on a milk container, you will notice that the milk has been homogenized. Homogenization involves mixing the various solutes so that they remain uniformly distributed in the milk; they do not settle to the bottom of the carton or bottle when left standing in the refrigerator. These solutes include proteins, carbohydrates, fats, vitamins, and minerals. The nutritional value of milk depends upon the vitamins and minerals that it contains. As you probably realize, milk is an important source of vitamin D, which is necessary for normal bone growth. In addition, milk contains several minerals required for good health. These include calcium, magnesium, zinc, phosphorus, and iron. In the following experiment, you can determine the amount of minerals present in milk. To check the mineral content, you will heat a milk sample until the solvent (water) evaporates, leaving the solutes (primarily the minerals) as a solid mass that can be weighed. By comparing the weight of the mineral mass with that of the original milk sample, you can calculate the mineral content percentage of milk.

Mineral Content of Milk

1. Heat a large, clean crucible with a Bunsen burner for five minutes to burn off any subtances present.
2. Remove the crucible with a pair of tongs and place it on an asbestos pad to cool. Transfer the crucible to a balance with the tongs and record its weight.
3. Fill the crucible approximately halfway with milk and then record the weight of the crucible and milk. *Gently* heat the crucible with a Bunsen burner (see Figure 2–1). Continue heating until all the milk has evaporated. After the milk has turned into a brownish-black ash, adjust the Bunsen burner to produce a more intense heat and continue heating for five minutes. The ash represents the mineral solutes that have remained after evaporating the solvent and other combustible substances.
4. Allow the crucible to cool and then weigh it along with the ash

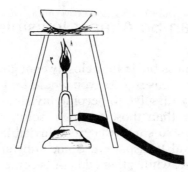

Figure 2–1. When evaporating the liquid portion of any solution, heat gently at first to avoid any splattering.

residue. By subtracting the weight of the crucible from that of the crucible and milk, determine the weight of the milk. By subtracting the weight of the crucible from that of the crucible and white ash, determine the weight of the ash. Calculate the mineral content of the milk by placing your data in the following equation:

$$\frac{\text{weight of ash (grams)}}{\text{weight of milk}} \times 100 = \text{percentage of mineral content}$$
$$\text{(grams)}$$

What assumption are you making about the solid ash material in this experiment?

Dilute and Concentrated Solutions

Depending upon the amount of solutes present, solutions can be dilute or concentrated. Given equal volumes of solvent, a concentrated solution contains more solutes than does a dilute one. You may wish to follow the above procedure using evaporated milk to determine if it is more concentrated than whole milk, at least with respect to their mineral contents. Based on your analysis, can you explain how a solution of evaporated milk is prepared? After comparing the mineral contents of whole and evaporated milk, can you give two ways in which any dilute solution could be made more concentrated? What about making a concentrated solution more dilute? What do evaporated milk and concentrated lemonade have in common?

Solutes Can Be Almost Invisible

If it were possible to look closely enough into a glass of milk to examine the various solutes, you would find that some of them do not completely dissolve to become invisible. Instead, these particles are larger than those found in solutions. In a solution, the solute becomes so small that it is invisible, usually making the liquid clear and transparent. Some substances in milk do completely dissolve, while others do not become small enough to form a true solution. Although they are too small to be seen even with a microscope, these particles remain large enough to be evenly suspended or spread throughout the milk. Whenever two substances are mixed so that one becomes suspended, but not dissolved, in the other, a *colloid* has been prepared. Milk is classified as a colloid.

Other examples of colloids include butter, whipped cream, gelatin desserts, mustard, jelly, and ketchup. In each of these, substances are broken down into particles that are suspended in a liquid. For example, butter consists of water particles that are evenly spread in milk fats. Whipped cream is made by suspending air in heavy cream. Ketchup is prepared by crushing tomatoes, spices, and sugar and then suspending them in water and vinegar. Have you ever had a problem getting the ketchup out of the bottle? The suspended particles are so concentrated that they trap the water. By vigorously shaking the bottle, you cause the particles to move further apart, thus releasing the water. Only then can you pour the ketchup on your hamburger or French fries!

Colloids Can Get It Together!

Not only can the suspended particles be moved further apart but they also can be brought closer together within the colloid. If enough particles are clumped together, either the colloidal solution will turn into a solid or a visible mass of material will form within the liquid. Whenever you prepare a gelatin dessert or salad, you begin by suspending the powdered materials in water. Refrigerating the colloid causes the particles to clump or coagulate, forming a solid that can then be eaten. If the solid gelatin were heated, the particles would move apart, changing the colloid back into a liquid. By the way, another colloid that can solidify under certain conditions is your blood. If a vein or capillary is injured, blood will flow from the cut; the colloidal particles in the blood will coagulate to produce a clot that stops the bleeding.

Methods to clump the particles to form a solid within the colloidal solution include changing the temperature or adding salts or acids. The proteins contained in milk are suspended in the liquid and can be coagulated by adding a dilute acid. In the following experiment, you can use this procedure to determine the protein content of milk. In order to coagulate all the proteins in milk, you should first isolate and remove the fats. You must exercise caution in conducting this experiment since the solvent used to extract the fats is flammable. The proteins can then be completely extracted from the fat-free milk. After finishing this experiment, you can calculate both the fat and the protein content of milk.

The Fat and Protein Content of Milk

1. Determine the weight of 100 ml of whole milk. Place the milk sample in a separatory funnel and add 100 ml of hexane or some other appropriate solvent as directed by your teacher. Carefully follow your teacher's instructions in adding the solvent to the milk. You may wish to conduct this portion of the experiment in a ventilating hood; if one is not available, open the classroom windows and doors to help circulate the air to remove any unpleasant odors. In any case, DO NOT BREATHE the fumes from the solvent. Do not use Bunsen burners or light matches, since hexane is flammable. Stopper the separatory funnel and gently shake the two liquids for a couple of minutes. The solution you added will extract the fats from the milk.
2. Allow the funnel to remain undisturbed until the two liquid layers are clearly separated (see Figure 2–2). Remove the stopper and slowly open the stopcock on the separatory funnel to collect the fat-free milk layer in a clean, dry beaker that has been weighed. The other liquid layer, which should be clearer, contains the fats that have been extracted from the milk. Save this liquid for possible use in the next step of this experiment.
3. Determine the weight of the fat-free milk sample by subtracting the weight of the empty beaker from that of the beaker and milk. The milk sample should weigh less than it did at the start of the experiment since the fats have been removed. If it does not, again mix this milk sample and solvent solution containing the fats you saved from step 2. Pour the mixture back into the

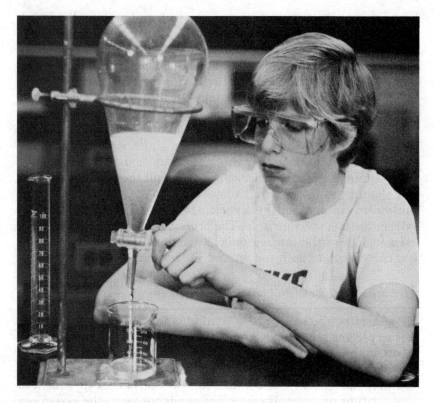

Figure 2–2. Allow sufficient time for the two layers to separate before opening the stopcock on the separatory funnel.

separatory funnel and repeat the extraction procedure. Once you have successfully extracted the fats, save the fat-free milk sample for use in determining the protein content as described in step 5. You can discard the solvent layer containing the fats.

4. Subtract the weight of the milk sample after removing the fats from the original weight of the milk. The difference represents the weight of the fat. Determine the percentage of fat in milk by placing your data in the following equation:

$$\frac{\text{weight of fat (grams)}}{\text{weight of milk sample (grams)}} \times 100 = \text{percentage fat}$$

5. To the fat-free milk sample that you saved from step 3, slowly add vinegar while stirring until solid clumps appear in the liquid. Any solid material formed in a solution is called a *precipitate*. The vinegar is a dilute acid solution that will cause the proteins in milk to coagulate into a precipitate.

6. To isolate the precipitate, pass the milk sample through a piece of filter paper that has been weighed. Allow the paper and precipitate to dry overnight and weigh them the next day. By subtracting the weight of the paper from that of the paper and precipitate, determine the weight of the proteins. Calculate the percentage of protein in milk by placing your data in the following equation:

$$\frac{\text{weight of protein (grams)}}{\text{weight of milk sample (grams)}} \times 100 = \text{percentage protein}$$

Let's Get All the Solutes Together at Once

As you have seen in the preceding experiments, you can isolate or coagulate each of the various substances present in milk. However, it is also possible to clump all the solutes and suspended particles at the same time. Made from the solid portion of milk, both cheese and cream cheese are commercially prepared in this manner. The solid portion is referred to as the *curd*, while the remaining liquid is called the *whey*. Before separating the curd from the whey, the milk must first be treated so that a chemical change takes place. Milk contains a sugar called lactose, which must be turned into another chemical substance, lactic acid. If milk is stored for a long enough time, the lactose will eventually change into lactic acid, causing the milk to sour. This chemical change, however, takes too long for the commercial preparation of cheese. To speed up this process, dairies add bacteria to the milk. The bacteria contain the chemical substances needed to turn lactose into lactic acid.

To prepare cheese in the following exercise, you will begin by adding buttermilk, which contains the chemicals required to change the lactose. To prepare the mixture, add 1 ml of buttermilk to 200 ml of regular milk. Since the sample must stand for at least four hours, you may want to prepare this mixture the day before you begin making the cheese.

Making Cheese—The Easy Whey

1. Place 200 ml of the milk sample in a large beaker. *Gently* warm the mixture to 30 to 32 degrees Celsius. Be careful not to overheat the milk. While stirring, slowly add five drops of rennilase

to the milk. If rennilase is not available, you can use a half tablet of rennin, which is sold in stores under the trade name Junket.

2. Slowly increase the temperature to 38 degrees Celsius and continue heating for several minutes. Be certain that the temperature does not get higher than 38 degrees. Discontinue heating and allow the beaker to cool. Examine the contents of the beaker and record your observations.

3. Filter the contents of the beaker through either filter paper or cheesecloth. Allow the solid material to dry. If you are courageous enough, add some salt and taste the cheese you have just prepared. How does it taste? To improve the flavor, commercial cheese is aged at between 35 and 55 degrees Celsius from several weeks to a couple of months.

Two Substances That Never Get Together

Have you ever tried to make a solution by mixing oil with either water or vinegar? No matter how hard you shake the two liquids, once you stop, the oil collects as tiny drops which eventually come together and separate from the water or vinegar. Obviously, oil does not form a solution or colloid with either of these liquids. However, the oil drops can be prevented from coming together by adding a third substance that stops the oil from separating from the other liquid. Whenever two liquids are mixed so that one remains distributed as small drops in the other, an *emulsion* has been prepared (see Figure 2–3). In order to maintain an emulsion, a third substance, known as an *emulsifier*, is needed to keep the drops of one liquid spread evenly throughout the other. Without the emulsifier, the drops would collect, forming a separate liquid layer. Use the following short recipe to prepare an emulsion that is found in every kitchen.

1. Mix one egg, 1 tablespoon of vinegar, ½ teaspoon of salt, ½ teaspoon of mustard and a ¼ cup of salad oil in a blender. Blend this mixture for several seconds.

2. Add ¾ cup of salad oil to the mixture in a slow, steady stream while blending until most of the oil is absorbed to form a uniform emulsion. Identify this emulsion. You may try varying these proportions or substituting different ingredients. For example, will lemon juice work in place of vinegar? What is the

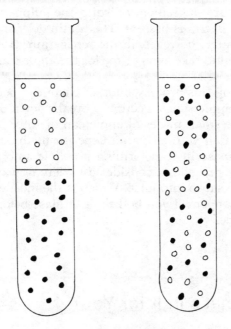

Figure 2–3. An emulsion is prepared when two immiscible liquids are spread evenly throughout each other.

purpose of adding the egg? Check various cookbooks for other sauces that require eggs.

From Colloid to Solution in One Easy Step

As you now realize, a close look in your refrigerator will reveal solutions, colloids, and emulsions. Move aside the solution, look behind the colloid near the emulsion, and you might find a colloid that can be turned into a solution—especially if you have a younger brother or sister. One such colloid is apple sauce, which usually consists of sugar, spices, and very small pieces of apple suspended in water. In preparing apple sauce, a chemical substance called pectin may be added to prevent the pieces of apple from getting even smaller. If they become so small and dissolve in the liquid, apple sauce (a colloid) would turn into apple juice (a solution)! To make apple sauce, the fruit is crushed into a pulp. Since pectin is a natural ingredient of fruits, the pulp remains a colloid as long as the pectin is present. The pectin does break down as it reacts with water; however, in the refrigerator this process is so slow that it would take a long time for the apple sauce to turn into apple juice.

To speed up this process, another chemical, pectinase, can be added to the apple sauce. Pectinase breaks down pectin; without the pectin, the small pieces of apple can dissolve in the water to form a juice. Can you design an experiment to demonstrate the effect of pectinase in transforming a colloid into a solution? In planning your procedure, consider a variety of ways in which to turn the apple sauce into juice. Which is the most effective procedure? What happens if you boil the pectinase before adding it to the apple sauce?

Juice—What's in It for You?

As a result of preparing apple juice, you now recognize that fruit juices are solutions containing very small particles dissolved in water. For example, orange juice contains several solutes including sugars, minerals, and vitamins. Some of the minerals are iron, calcium, and phosphorus. The vitamins include vitamins A, B, and C. Also known as ascorbic acid, vitamin C is required for good health. Used for the repair of injured tissue, vitamin C is also involved in many chemical reactions needed to keep your body functioning normally. Since your body cannot produce vitamin C, you must obtain it from foods, liquids, or vitamin tablets. As you probably realize, fruit juices have a higher percentage of vitamins than soft drinks. But what about fruit drinks or fruit sodas, especially those advertised as having extra vitamin C? Are they just as good for you as fruit juices? One way to find out is to analyze these different fruit solutions to determine how much vitamins and minerals they contain. In the following experiment, you can compare the levels of vitamin C in orange juice, drink, and soda.

Testing for Vitamin C

1. Before analyzing various solutions for their vitamin C content, you must establish a standard to be used as a reference point when performing your calculations. Add a crushed 100 mg vita-

min C tablet to 100 ml of distilled water. After mixing the tablet and water thoroughly, pour 50 ml of the solution into a clean, dry flask. Add 2.5 ml of 1 percent soluble starch solution to the flask. This is your standard solution containing 50 milligrams of vitamin C. Fill a buret or graduated cylinder with Lugol's iodine solution and record the volume of iodine.

2. *Slowly* add the iodine drop by drop to the flask containing the vitamin C and starch (see Figure 2–4). Swirl the flask after adding each drop of iodine. A blue-black color may form but will disappear upon swirling the contents of the flask. Continue adding the iodine until a blue-black color persists for at least one minute after swirling the flask. Record the volume of iodine

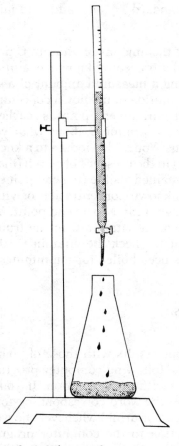

Figure 2–4. When performing any titration, swirl the contents of the flask or beaker after the addition of each drop.

used to reach this point. At this stage, all the vitamin C (50 mg) has reacted with the iodine; the starch can then combine with the iodine, forming the blue-black color that remains visible.

3. Pour 50 ml of orange juice, soda, or drink into a clean flask and add 2.5 ml of the 1 percent soluble starch solution. Following the same procedure as before, add the Lugol's iodine until the blue-black color remains visible. Record the volume of iodine added to each solution tested.

4. Calculate the amount of vitamin C present in the solution by placing your data in the following equation:

$$\frac{50 \text{ mg (vitamin C standard)}}{\substack{\text{volume of iodine}\\\text{added to standard}}} = \frac{X \text{ (juice solution)}}{\substack{\text{volume of iodine}\\\text{added to juice}}}$$

In determining the amount of vitamin C present in the solution, you performed a procedure known as a *titration*. Titration is the process of adding a measured amount of a solution of known concentration to a solution of unknown concentration in the presence of an indicator until an end point is reached. Sound complicated? If you think for a minute about what you did, you'll see what titration means. You added iodine to a known concentration of vitamin C (50 mg) in the presence of starch (the indicator) until a blue-black color remained visible (the end point). You then added the iodine to an unknown concentration of vitamin C (the fruit solution) until you again came to an end point. After titrating, you calculated the amount of vitamin C in the fruit solution. By the way, you may want to check the vitamin C content of the fruit solution after it has been boiled for ten minutes.

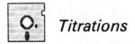 ## Titrations

To compare your results with those of your classmates, you may want to use the following computer program. Type it in exactly as it appears, with one exception. If you did not analyze orange juice, type in the word corresponding to the solution that you tested (either soda or drink) whenever the word *juice* appears. Check the introduction to the computer program in Chapter 1 (pages 6–7) for screen-clearing commands. When the following program is run, it will give you the average amount of vitamin C based on the number of titrations conducted.

```
10 PRINT "HOW MANY TITRATIONS WERE PERFORMED"
20 PRINT "TO TEST THE AMOUNT OF VITAMIN C"
30 PRINT "IN THE ORANGE JUICE SAMPLE ";
40 INPUT N
50 PRINT "ENTER THE NUMBER OF MILLILITERS OF"
60 PRINT "IODINE ADDED TO THE ORANGE JUICE"
70 PRINT "FOR EACH TITRATION."
80 FOR X = 1 TO N
90 INPUT A
100 B = B + A
110 NEXT X
120 PRINT "HOW MANY TITRATIONS WERE PERFORMED"
130 PRINT "TO TEST THE STANDARD VITAMIN C"
140 PRINT "SOLUTION ";
150 INPUT L
160 PRINT "ENTER THE NUMBER OF MILLILITERS OF"
170 PRINT "IODINE ADDED TO THE STANDARD"
180 PRINT "VITAMIN C SOLUTION."
190 FOR X = 1 TO L
200 INPUT S
210 T = T + S
220 NEXT X
230 P = T/L:Q = B/N
240 J = (50*Q)/P
250 PRINT "THE AVERAGE AMOUNT OF VITAMIN C"
260 PRINT "IN THE ORANGE JUICE SOLUTION IS"
270 PRINT J;" MILLIGRAMS."
```

Gas a Solution and What Do You Get?

If you again look in the refrigerator, there is a solution you will easily identify—a soft drink. No matter whether it is called soda, soda pop, or just plain pop, all soft drinks basically contain the same solutes dissolved in water. These include a sweet syrup, artificial flavorings, and carbon dioxide gas to give it a bubbly taste. Have you ever shaken a soft drink bottle or can and then quickly removed the cap? The soda sprayed out from the bottle as soon as you took off the cap. To make the soft drink, the manufacturer adds carbon dioxide gas under pressure and then carefully seals the bottle or can to prevent the gas from escaping. If you shake up the container, the gas bubbles come together, creating a strong pressure that forces out the water in a spray when you open the seal. You can see these gas bubbles come to the surface of the liquid whenever you open a soft drink. If you leave the open container of soda standing for a long enough time, the drink will taste flat because most of the carbon dioxide gas has escaped from the water. The gas is responsible for the sparkling taste that is characteristic of carbonated solutions.

Although all carbonated beverages contain carbon dioxide gas, one solute that may differ in various types of soft drinks is sugar. Whereas regular drinks contain sugar as the main sweetening ingredient, diet beverages use a substitute such as saccharin. In using a sugar substitute, only a small amount is added since saccharin is 300 times sweeter than sugar. By replacing sugar with saccharin, the manufacturer is reducing the number of calories in the soft drink for those who are concerned about their weight. If diet beverages contain less sugar, how would their density compare to that of regular soft drinks? Recall that density represents the amount of mass present in a given volume. Expressed as a formula, density is equal to mass (usually expressed in grams) divided by the volume. In the following experiment, you can compare the densities of regular and diet soft drinks.

Regular Versus Diet Soda—Who's Denser?

1. Record the weight of a clean, dry 100-ml cylinder. Add 100 ml of a regular soft drink to the cylinder. You must be as accurate as possible. If you wish, use a dropper to obtain 100 ml exactly; also be sure to read the bottom of the *meniscus*, the curvature formed by a liquid in a narrow container (see Figure 2–5).
2. Weigh the cylinder and beverage. Calculate the density of the regular soda by dividing the weight of the beverage by 100 (the volume of the liquid).
3. Repeat this procedure using a diet soft drink.
4. Explain your results. How could you increase the density of the diet soft drink? What could you do to decrease the density of the regular beverage?

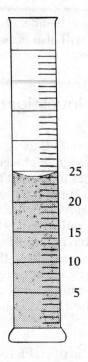

Figure 2–5. What is the volume of liquid in this graduated cylinder?

How Sweet It Is!

Since they have a low concentration of vitamins and minerals, soft drinks are not a good source of nutrients necessary for good health. Unlike orange juice, sodas have a high ratio of sugar to vitamins and minerals. Although they may quench your thirst on a hot summer day, soft drinks are not very beneficial to your health. If you drink several soft drinks a day, you are gulping down large amounts of sugar without getting any other nutritional value. To test different soft drinks for the presence of sugar, perform the following experiment.

Place 5 ml of the soft drink and 3 ml of quantitative Benedict's solution in a test tube. Place the test tube in a boiling-water bath for five minutes. The Benedict's solution is used to test for certain sugars. If the solution in the test tube turns orange or red after heating, then sugar is present. The darker the color, the more sugar present. You can use Benedict's solution to test any liquid substance for the presence of certain sugars. By the way, the next time you're thirsty, try the following recipe: mix equal volumes of orange juice and carbonated water (soda or club water), add some crushed ice, and mix it in a blender. Not only will this solution quench your thirst, but it will also provide the minerals and vitamins missing in soft drinks.

Topics for Further Investigation

The importance of vitamin C for good health has been known for some time. As early as the eighteenth century, British ships were required to carry limes to prevent a disease known as scurvy. Like all citrus fruits, limes contain large amounts of the vitamin C needed to prevent this disease. Because they ate limes as a regular part of their diet, British sailors are referred to as "limeys." Recently, published reports have claimed that vitamin C is also helpful in preventing other diseases—from the common cold to cancer! Other experiments have failed to demonstrate any such connection between vitamin C and the prevention of disease. Why should scientists disagree? Check the claims issued by both sides. What is your opinion?

In addition to sugars, artificial flavorings and carbon dioxide gas, soft drinks may also contain caffeine. As you probably know, caffeine functions as a stimulant to help keep you awake. However, concerns have been raised about the possible side effects of too much caffeine. Examine the chemical nature of caffeine in more detail and discuss the concerns that have been raised about the potential dangers of overconsumption of caffeine.

Although milk is very healthy for you, it can cause severe problems in a small percentage of newborn babies. How is this so, and can anything be done to prevent a problem from arising?

Milk contains many nutrients and is used as the main source of nourishment for newborn mammals. Does the milk produced by different mammals, including humans, dogs, cats, cows, goats, and rabbits, have the same chemical composition? How are milk substitutes prepared?

The existence of colloids is important in a variety of organisms—from amoeba to man. Examine the function of colloids in different organisms, specifically the formation of blood clots in humans. The inability to stop bleeding is known as hemophilia. What major problems exist in the treatment and cure of hemophiliacs?

Suppose you were placed in charge of developing a drink to be used by all athletes training for the Olympics. What would your solution contain? Defend the basis for selecting each ingredient.

*** Research chemists are constantly searching for artificial sweeteners that taste just like glucose but without the calories. Although products have been synthesized to do just that, studies have raised questions about the potential health hazards caused by some of these products. You may be interested in developing an artificial sweetener that contains few calories and poses no health hazard.

*** Crystal formation is an interesting phenomenon, but much remains to be learned about the chemistry involved. Crystals can be formed in supersaturated solutions, that is, those containing more dissolved solutes than they normally hold at a given temperature. Examine factors that influence the formation of crystals in supersaturated solutions; this might include investigating the chemical nature of the solute, the type of solvent, and environmental factors such as pressure.

3 Food—Packaged Energy

How many times have you been told to eat well-balanced meals? Most likely someone, either your parents or a teacher, has often reminded you that your meals should supply the nutrients needed for good health, strength, and growth. One such dinner would include liver for its protein, vitamins, and minerals; spinach for additional vitamins; and rice for its starch. Chances are, however, that you don't eat this meal very often. In fact, sometimes we just don't have the time to eat any type of healthful and well-balanced meal. Instead, we grab a doughnut at home for breakfast, a cheeseburger and french fries at a fast-food restaurant for lunch, or hot dogs and potato chips at a football game for dinner. Although these foods satisfy our hunger, they provide only limited quantities of a few important nutrients needed for good health. Moreover, if we maintained a steady diet of these foods, the extra pounds would quickly begin to show.

As you probably know, the increased weight would result from the large number of calories present in these foods. A *calorie* is the unit scientists use to measure the energy content of foods. French fries, potato chips, and doughnuts are examples of foods with a high calorie content. Consequently, these foods constitute a rich energy source for our bodies. If eaten as a regular diet, however, these foods provide more calories than we can normally use for moving our muscles, keeping warm, or building new cells for growth. As a result, many unused digested food nutrients, along with their calories, are stored in the cells and therefore add to the body's weight. A well-balanced meal, on the other hand, not only limits the number of calories but also supplies the minerals and vitamins missing in so-called junk foods.

One type of junk food that you may eat whenever you want a quick burst of energy is a candy bar. Believing that the chocolate bar would supply your body with a spurt of energy, you may have eaten one just before competing to provide that extra speed during the race or that stronger serve in tennis. Yet, there is little nutritional value in the candy bar. The chocolate, which is not sweet at all, is simply added for flavoring. The main ingredient is sugar, which gives the candy its characteristic sweet taste. Sugars are classified as *carbohydrates*, one of three categories under which all foods can be listed. The other two are *proteins* and *lipids*. The latter includes a group of chemical compounds which are probably more familiar to you—the fats.

When compared on an equal basis with either proteins or carbohydrates, fats contain more energy. Although this may be difficult to believe, it would make sense if you think about it in another way—fat contains more calories than an equivalent amount of protein or carbohydrate. For example, 100 grams of bacon, which is half fat, contains twice the number of calories found in the same amount of angel food cake, which is approximately half carbohydrate. But before you consider eating something with a high fat content to obtain some extra energy in your next race, you should realize that the cells of your body have more difficulty in releasing the energy from fats than from sugars. So for that quick burst of energy, sugars are a better bet.

To determine the amount of energy present in a food sample, a special instrument known as a *calorimeter* is used. The calorimeter measures the amount of energy released from a food sample in the form of heat. The food sample is placed in a combustion chamber and is then ignited by electrical sparks. A larger, insulated container filled with water surrounds the combustion chamber. After the food is ignited, all its energy is released as heat, which is absorbed by the water in the calorimeter. As the food burns, the temperature of this water is raised. Knowing the temperature increase and the weight of the water, scientists can then calculate the number of calories in a food sample. In the following experiment, you will learn how to build a simple calorimeter. Although the measurements will not be as accurate as those obtained with an insulated calorimeter, you will be able to determine and compare the calorie content of various food samples.

A Calorimeter—What Every Weight Watcher Needs

1. Using a large metal can, construct a calorimeter as shown in Figure 3–1. A coffee can or any other large, metal container would be suitable. After removing the top, cut out a pie-shaped

Figure 3–1. Once the food sample is ignited, avoid any small drafts that might extinguish the fire.

    piece along the side and invert the can so that the bottom is upright. Carefully cut a hole wide enough to hold a large test tube. Puncture several air holes around this opening.

2. Insert a pin into a small wooden cork and place this under the can. Put 10 ml of water into the test tube and position it so that the bottom is approximately 2 cm above the pin. Once you have positioned the test tube, remove the cork and pin from under the can. You are now ready to use the calorimeter to determine the calorie content of a food sample.

3. Weigh 0.2 gm of cured pork fat or any other food that is easy to burn, and insert the food into the pin.

4. Place a thermometer into the test tube and record the temperature of the water.

5. Ignite the food sample with a match; once it is burning, carefully place the cork and pin under the test tube in the calorimeter. If you have difficulty in igniting the food sample, you can use a very small piece of string as a wick. Mold the food sample around the string, and then ignite the string. You may have to

repeat this process several times before you are successful in igniting the food.

6. After the pork fat has completely burned, observe the thermometer and record the highest temperature of the water.
7. A calorie is defined as the amount of heat energy required to raise 1 gm of water one degree Celsius. To determine the number of calories in your food sample, insert your data in the following equation:

calories = volume × (final temperature − initial temperature)

What assumption are you making in using this formula? Explain why the calculations derived from your calorimeter are not as accurate as those obtained from an insulated, commercial calorimeter. After determining your answer, you were probably surprised by the rather large number of calories present in the small piece of food. Actually, there are two kinds of calories—the scientific ones and the ones used in diets. In calculating the calorie content in the preceding experiment, you were working with scientific calories, spelled with a small letter. Whenever a reference is made to diets, the word *Calorie* is written with a capital letter. One Calorie is equal to 1,000 calories. How many Calories did your food sample contain? You may wish to test other food samples to compare their calorie content. Check some references to find out what the recommended daily allowance of calories or Calories is for you.

Calories

If you are interested in checking your understanding of calories, you may want to type the following program into your computer. This program will allow you to select any three of the four experimental conditions in this investigation—amount of water used, initial and final water temperatures, and the number of calories. After selecting any three of these values, you will be asked to determine the fourth one. Can you set up the experimental conditions necessary to generate the number of calories recommended for your daily allowance? For example, what might be the initial and final temperatures of water if you used 100 ml in the calorimeter to produce this number of calories? If you wish to clear the screen after each run, check the introduction to the computer program in Chapter 1 (pages 6–7) for screen-clearing commands for the Apple II and TRS-80 systems.

```
5 PRINT "WHAT IS YOUR FIRST NAME";:INPUT NA$
10 PRINT "TYPE IN THE NUMBER THAT CORRESPONDS"
20 PRINT "TO THE VALUE YOU WISH TO CALCULATE."
30 PRINT "1. CALORIES"
40 PRINT "2. AMOUNT OF WATER"
50 PRINT "3. INITIAL TEMPERATURE OF WATER"
60 PRINT "4. FINAL TEMPERATURE OF WATER":PRINT
70 PRINT "WHICH NUMBER";:INPUT X
80 IF X = 1 THEN 300
90 IF X = 2 THEN 400
100 IF X = 3 THEN 500
110 PRINT
120 PRINT "HOW MANY CALORIES ARE PRODUCED";:INPUT C
130 PRINT "HOW MANY ML OF WATER ARE USED";:INPUT W
140 PRINT "INITIAL WATER TEMPERATURE";:INPUT I
150 PRINT
160 PRINT "WHAT IS THE FINAL WATER TEMPERATURE";:
    INPUT F
170 IF F = (C + (W*I))/W THEN 190
180 PRINT "THAT IS NOT CORRECT. TRY AGAIN, ";NA$;".":GOTO
    150
190 PRINT "VERY GOOD, ";NA$;", YOU ARE CORRECT!"
200 GOTO 600
300 PRINT
310 PRINT "HOW MANY ML OF WATER ARE USED";:INPUT W
320 PRINT "INITIAL WATER TEMPERATURE";:INPUT I
330 PRINT "FINAL WATER TEMPERATURE";:INPUT F
340 PRINT
350 PRINT "HOW MANY CALORIES ARE PRODUCED";:INPUT C
360 IF C = W*(F − I) THEN 380
370 PRINT "THAT IS NOT CORRECT. TRY AGAIN, ";
    NA$;".":GOTO 340
380 PRINT "VERY GOOD ";NA$;", YOU ARE CORRECT!"
390 GOTO 600
400 PRINT
410 PRINT "HOW MANY CALORIES ARE PRODUCED";:INPUT C
420 PRINT "INITIAL WATER TEMPERATURE";:INPUT I
430 PRINT "FINAL WATER TEMPERATURE";:INPUT F
440 PRINT
450 PRINT "HOW MANY ML OF WATER WERE USED";:INPUT W
460 PRINT
470 IF W = C/(F − I) THEN 490
480 PRINT "THAT IS NOT CORRECT. TRY AGAIN, ";
    NA$;".":GOTO 440
```

```
490 PRINT NA$;"', YOU ARE CORRECT!":GOTO 600
500 PRINT
510 PRINT "HOW MANY CALORIES ARE PRODUCED";:INPUT C
520 PRINT "HOW MANY ML OF WATER ARE USED";:INPUT W
530 PRINT "FINAL WATER TEMPERATURE";:INPUT F
540 PRINT
550 PRINT "WHAT IS THE INITIAL WATER"
560 PRINT "TEMPERATURE";:INPUT I
570 IF I = (W*F − C)/W THEN 590
580 PRINT "THAT IS NOT CORRECT. TRY AGAIN, ";NA$;".":GOTO
    540
590 PRINT NA$;"', YOU GOT THE CORRECT ANSWER."
600 PRINT
610 PRINT "DO YOU WISH TO TRY ANOTHER"
620 PRINT "PROBLEM, ";NA$;"(Y OR N)";:INPUT AN$
630 IF LEFT$(AN$,1) = "Y" THEN 10
640 END
```

FAT—WHAT EVERY WEIGHT WATCHER DOESN'T NEED

Actually, everyone needs some fat as a part of their regular diet to stay healthy. In conducting nutritional studies, scientists have discovered that a totally fat-free diet can lead to serious health problems. Since fats play an important role in the building of many organs, including the brain, liver, lungs, and kidneys, the absence of fats in a diet would prevent the normal development of these structures. In addition, you may recall, fats constitute a valuable source of energy for our daily needs.

On the other hand, too much fat in the diet could also affect your health. A disease known as cirrhosis is caused by the accumulation of fat in the liver. These fat deposits can damage liver cells; if the level of fat continues to increase, these cells may die. Consequently, the liver will no longer be able to perform one of its many vital functions—the removal of toxic chemicals from the body. If these poisonous substances are not eliminated, serious illness, and possibly death, can result.

Approximately 20 to 25 percent of your daily caloric intake represents the proper balance between too little and too much fat in your diet (see Figure 3–2). Fat is present in most foods, including bacon, cheese, peanuts, potato chips, margarine, and butter.

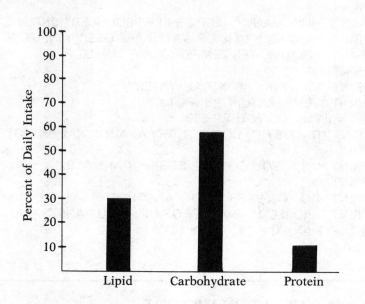

Figure 3–2. According to U.S. government figures, nearly two-thirds of your daily food intake should consist of carbohydrates.

These fats are solid at room temperature; those that are liquids at room temperature are called *oils*. These include oils from a variety of vegetables, including peanuts, coconuts, olives, and corn.

Fats and oils have many common characteristics; for example, neither mixes with water but both dissolve in solvents such as hexane, chloroform, and dichloromethane. You may have used one of these solvents to extract the fats from milk in Chapter 2. There is, however, one very important chemical difference between most fats and oils, a difference that can have a significant effect on your health. Because of their chemical structure, oils are less likely to cause problems with the circulation of blood through your heart and blood vessels. Consequently, vegetable oils are recommended over fats for dietary purposes, including cooking. To understand the difference between the chemical composition of fats and oils, you must first examine the chemical structure of lipids in general. If you do not wish to study the chemistry of lipids in greater detail, you can proceed directly to the section titled "Tasty or Greasy."

Better Living Through Organic Chemistry

Like proteins and carbohydrates, lipids are classified as *organic* compounds. You may be familiar with the word organic from having heard the phrases "organically grown" or "contains only natural, organic ingredients." Originally, scientists used the term "organic" to describe any chemical compound found in a living organism. Those compounds limited to nonliving systems, such as rocks and water, were called *inorganic*. Today, however, scientists no longer use this criterion to distinguish between the two. Instead, an organic compound is defined as any chemical substance that contains the element carbon. *Elements* are the building blocks of chemical compounds.

A few compounds containing the element carbon are not classified as organic; for example, carbon monoxide and carbon dioxide are considered inorganic compounds. In addition to carbon, organic compounds often contain the elements oxygen and hydrogen. Organic compounds are now recognized to be part of both living and nonliving systems. In fact, scientists are constantly searching for ways to combine carbon, hydrogen, and oxygen to make new organic compounds. These man-made substances include medicines, plastics, perfumes, clothing materials, insecticides, and fertilizers. Obviously, organic compounds play an important role in our daily lives.

As one of the major classes of organic compounds in foods, lipids are characterized by having many more carbon and hydrogen atoms than oxygen atoms. For example, examine the following chemical formula for a lipid: $C_{57}H_{110}O_6$. A *formula* is a shorthand method for providing some information about a chemical compound. This formula tells us that there are 57 carbon atoms, 110 hydrogen atoms, and only 6 oxygen atoms in this compound. An *atom* is the smallest particle of an element that can combine with another element. If you examine the formula again, you will notice that the proportion of carbon and hydrogen to oxygen is very high; this is a chemical characteristic of lipids.

A chemical formula, however, does not provide enough information to show the important difference between fats and oils. In order to understand this difference, you must examine another type of formula, called a *structural formula*. A structural formula shows the arrangement of all the atoms of the various elements in a compound. For example, the structural formula for a chemical compound found in butter is:

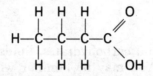

Each atom is represented by a letter—C for carbon, H for hydrogen, O for oxygen. Notice that these atoms are connected to one another, as illustrated by the lines in the structural formula. Compare the above structural formula with the following one, which represents a chemical substance found in vegetable oil:

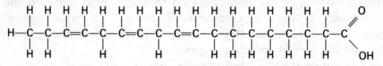

Is there any similarity between the organic compound present in butter and the compound found in the vegetable oil? As lipids, both compounds contain a higher number of carbon and hydrogen atoms than oxygen atoms. Is there any noticeable difference? If you examine the lines connecting several of the carbon atoms, you will see that double lines are present in the vegetable oil in three places; no double lines are drawn between carbon atoms in the butter fat. These lines, whether single or double, represent *chemical bonds*, which connect or join one atom to another. Double bonds between carbon atoms are present in the vegetable oil but not in the butter fat. Whenever a lipid contains one or more double bonds connecting carbon atoms, it is said to be *unsaturated*. If all these bonds are single ones, the lipid is said to be *saturated*.

Molecular model kits may be helpful in demonstrating the chemical difference between unsaturated and saturated fats. If they are available, use the kits to construct models of different fats. Can you build a fat that is highly unsaturated? How many double bonds are present? Check a reference source for the formulas of various fats. Can you build these compounds, using the molecular model kits?

Be Wary of Saturated Fats

Research reports have indicated that saturated fats raise the level of another chemical compound, *cholesterol*, in the blood. Too much blood cholesterol is known to cause hardening of the arteries, preventing the smooth flow of blood through the arteries and veins and possibly increasing the risk of a heart attack. For this reason, unsaturated fats, found in most vegetable oils, are recommended over saturated fats for diets. To reduce their intake of saturated fats, many families have switched from butter to margarine. Butter contains approximately 97 percent saturated fat; margarine, only 60 percent saturated fat. "Soft" margarine, which spreads more easily, is even lower in saturated fat and therefore may be more effective in controlling blood cholesterol levels.

You can use the following test to determine whether a fat is saturated or unsaturated. Unsaturated fats will undergo a chemical reaction when the double bonds between the carbon atoms are broken, making room for more atoms. Additional atoms can then be added to an unsaturated fat. For example, iodine atoms can be added to an unsaturated fat, as illustrated in the following diagram:

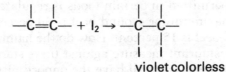

Notice that the iodine, which has a violet color, becomes colorless when added to the unsaturated fat. The degree of unsaturation of various fats can also be determined. The more unsaturated the fat, the greater the number of double bonds between carbon atoms. Since more double bonds are available to be broken in a highly unsaturated fat, more iodine can be added until all the double bonds have been changed into single ones. At that point, any additional iodine would not react and would consequently remain violet. Iodine cannot be added to a saturated fat since there are no double bonds that can be broken. Therefore, a few drops of the iodine solution should produce a violet color when added to a saturated fat.

Saturated or Unsaturated? That Is the Question

Mix 5 ml of hexane or another appropriate solvent with 3 ml of oil in a test tube. Do not use the hexane near any flames since it is very flammable. After stirring the solvent and oil for several minutes, add one drop of 0.05 percent iodine solution, made by dissolving 0.05 gm of iodine in 100 ml of the solvent you are using. Continue adding the iodine drop by drop until a violet color remains clearly visible. Stir the contents of the test tube after adding each drop. Hold the test tube against a sheet of white paper to observe the color. Record the number of drops added to each of the oil samples you are testing. Which of the oils has the higher percentage of unsaturated fats? Explain why the color remains visible after a certain point is reached.

Tasty or Greasy?

Foods taste better because of the fats they contain. However, too much fat is not only unhealthy but also responsible for giving food a greasy taste. How many times have you eaten at a fast-food restaurant where the French fries were greasy? You may have even noticed that they left grease spots on the paper container. The amount of fat permitted in certain foods is regulated by law. For example, the maximum for ground beef is 30 percent, while that for lean ground beef is 15 percent. How do the hamburgers served at a fast-food restaurant measure against these standards? In the following experiment, you will have the opportunity to check the fat content of the ground beef used to make hamburgers.

How to Spot a Fat Hamburger

Since the fat content should be determined before the meat is cooked, try to obtain the ground beef used for making hamburgers at a fast-food restaurant before they are prepared. Explain that you are conducting a scientific investigation requiring uncooked hamburger meat. If you cannot obtain any hamburger patties before they are cooked, you can use different types of ground meat available from grocery stores, including sirloin, round, chuck, and beef. Refrigerate some of the ground meat for later analysis, including tests for starch and protein. If you record the prices, you can determine which one is the best buy for your money—which provides the most meat with the least fat at the best price.

1. Weigh 25 gm of the uncooked, ground meat and place it in a large beaker.
2. Add 25 ml of hexane or another appropriate solvent to the meat and mix thoroughly for 15 to 20 minutes with a glass stirring rod. Do not use the hexane near any flames since it is very flammable. The solvent will extract the fat from the meat.
3. Record the weight of the bottom half of a petri dish or any

container large enough to hold 25 ml and wide enough to allow for evaporation of the solvent. Carefully drain the liquid into the petri dish or container.

4. Allow the solvent to evaporate overnight. The next day, weigh the dish and fat. By subtracting the weight of the dish from that of the dish and fat, calculate the amount of fat. Determine the percent of fat in the meat sample by placing your answer in the following equation:

$$\frac{\text{weight of fat}}{25 \text{ gm (meat sample)}} \times 100 = \text{percent fat}$$

5. Which brand of fast-food hamburger has the least fat? Is it worth paying the difference in price at a grocery store for ground sirloin?

6. While you have the meat samples, you may want to analyze them for their water content. Water is sometimes added to meat to make it juicy. Can you design an experiment to determine the water content of the different meat samples? Knowing the boiling point of water may be helpful in planning your procedure. What would happen to the meat if it were placed at this temperature? What other boiling points should you check in a chemistry reference book before making conclusions about the water content? You may also find it interesting to compare different brands of hot dogs for their fat and water content. You should use half a hot dog for each analysis. Which brand of hot dog is the best buy? Which brand can be advertised as a "juicy frank"?

CARBOHYDRATES—POWER FROM PLANTS

Using solar energy in the process of *photosynthesis*, green plants convert two inorganic compounds, water and carbon dioxide, into carbohydrates, one of the major classes of organic compounds in the foods you eat. Carbohydrates are the main energy compounds in your diet. Although an equivalent amount of fat contains more calories, carbohydrates represent an energy source that is easier for your body cells to use.

As organic compounds, carbohydrates contain the elements carbon, hydrogen, and oxygen. Carbohydrates, which include sugars and starches, are characterized by having two hydrogen atoms for each oxygen atom. For example, the formula for a simple carbohydrate, glucose, is $C_6H_{12}O_6$. If you examine this formula, you will notice that the ratio of hydrogen to oxygen in carbohydrates is 2:1 or the same as that in water, whose formula is H_2O. How do carbohydrates differ chemically from lipids? How are they similar?

Glucose is the only carbohydrate your body can use for its energy requirements. All other sugars and starches in your diet must first be converted to glucose before they can be used for energy. You may be interested in analyzing different food products to see if they contain glucose. To test for glucose, use the Benedict's solution described in the section on soft drinks in Chapter 2. You can test both dry and liquid samples. When testing a liquid, mix 5 ml with 3 ml of Benedict's solution (see Figure 3–3). For solids, first boil a small sample in 5 ml of water and then test this solution with Benedict's. If you have quantitative Benedict's solution, you can check to see which food product has the highest level of glucose.

The Taste Test

Two other examples of sugars are fructose and galactose. In fact, these sugars have the same formula as glucose. Would you expect them to be identical chemical compounds with different names? To find out, try the following taste test!

Place a small amount of galactose on your tongue. Does it taste sweet, as you would expect of a sugar? Rinse your mouth with water, and then taste a small sample of glucose. Rinse again, and test fructose. Rank the sugars in order of their sweetness. If they have the same formula, how can there be a chemical difference between them? You may want to check their structural formulas. By the way, *never* taste chemicals unless you are absolutely sure that they are harmless and safe to place on your tongue!

Are Sugars Really That Simple?

Glucose, galactose, and fructose are classified as simple sugars, or *monosaccharides*. Simple sugars cannot be broken down into other sugars. They can, however, be combined with one another to form more complex sugars. For example, glucose and fructose can be joined to form sucrose, which is the sugar you put in

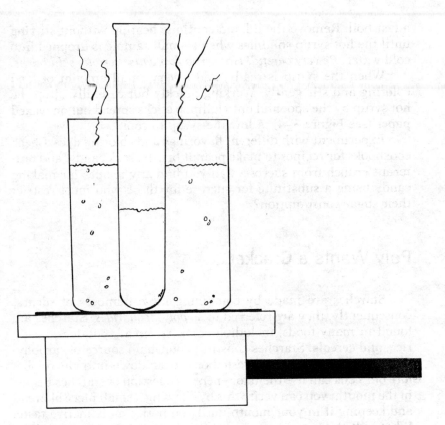

Figure 3–3. A positive Benedict's test for sugar will be indicated by a color change from blue to green, orange, or red. When analyzing for sugar, be sure the level of the water in the beaker is above that of the solution in the test tube. The solution must be heated in a boiling-water bath for five minutes.

coffee and use in cooking. Since sucrose is composed of two sugars, it is referred to as a *disaccharide*. Maltose, used for making milkshakes, and lactose, the sugar in milk, are examples of other disaccharides commonly found in a kitchen. Many candies are made from disaccharides, especially sucrose. In the following experiment, you can try your hand at being a candy maker!

How to Make Lollipops

To make approximately 20 lollipops, combine 4 cups of sugar, 1 ⅓ cups of light syrup, 2 cups of water, and ¼ teaspoon of salt in a large cooking pot. Stir well, cover tightly with a lid, and bring to a fast boil. Remove the lid and continue heating without stirring until the hot syrup solidifies when a small sample is dropped into cold water. *Be very careful! Hot syrup can cause serious skin burns.*

When the syrup is ready, add a very small amount of food coloring and stir gently. Working quickly but carefully, drop the hot syrup by the spoonful onto lollipop sticks spread out on waxed paper (see Figure 3–4). Allow the syrup to cool.

Experiment with different flavorings and coloring dyes. Check cookbooks for recipes to make peanut brittle, rock candy, and butternut crunch from sucrose. Can you find any recipes for making candy using a substitute for sucrose for those who must restrict their sugar consumption?

Poly Wants a Cracker

Starches are made by combining large numbers of sugars; consequently, they are referred to as *polysaccharides*. Starches are found in many foods, including crackers, bread, potatoes, pasta, rice, and cereals. Starches constitute our main source of carbohydrate, but they must be digested or broken down into glucose before our cells can use them for energy. Digestion of starches begins in the mouth; you can verify this by chewing a small piece of bread and keeping it in your mouth until you notice a distinctive taste. What will the bread taste like if you keep it in your mouth long enough?

The presence of starch can be determined by adding a few drops of Lugol's iodine solution to 2 ml of the sample to be tested. The appearance of a blue-black color indicates the presence of starch. You may find it interesting to test various food products for starch. In the case of solid samples, boil a few grams of the food in a small amount of water and then test the liquid for starch. Starch is occasionally added to foods, including some hamburger meat and hot dogs, as a "filler." Do any of your fast-food hamburgers contain starch as a filler?

Other common polysaccharides include glycogen and cellulose. Known as animal starch, glycogen is stored in the liver and muscles as an energy reserve. When needed, glycogen is released into the bloodstream, where it can be taken in by cells and broken down into glucose. Cellulose is found primarily in plants, where it provides the structural support needed to keep the plant upright. Many vegetables are rich in cellulose. Although it is part of our regular diet, cellulose cannot be broken down into glucose by the

Figure 3–4. The hot syrup can be poured into a beaker to cool. Before it solidifies, the syrup can then be poured onto wax paper.

human digestive system to supply energy. Consequently, cellulose is eliminated from the body as waste; its sole function is to provide the roughage needed to prevent constipation.

Cooking vegetables softens the cellulose by changing it into smaller sugars, making it easier to pass through our digestive system. Try cooking a vegetable high in cellulose, such as squash or eggplant, in three different ways—in water, in baking soda, and in vinegar. Taste each sample to see which is the softest and therefore the one broken down to the greatest extent into simpler sugars.

PROTEINS—THE COMPOUNDS OF LIFE

Have you ever heard the expression, "You are what you eat"? The truth of this saying is apparent when you consider the role proteins play in our daily lives. *Proteins*, the third major class of organic compounds in foods, are used in building many parts of our bodies. Nearly half of the protein we consume in our diets is used in building structures such as muscles, blood vessels, hair, and skin. Every day we lose thousands of skin cells, which must be replaced. Proteins play a vital part in this rebuilding process. In addition, proteins are also used to build other compounds needed to regulate the chemical reactions taking place in our bodies. These compounds include *enzymes* and *hormones*. Enzymes speed up chemical reactions taking place in our bodies, including those involved in digestion. Without enzymes, our cells would be unable to obtain the nutrients needed for energy, growth, and repair. Hormones serve as chemical regulators, controlling the levels of other chemical compounds in the bloodstream.

As organic compounds, proteins contain carbon, hydrogen, and oxygen. Unlike lipids and carbohydrates, proteins contain another element, nitrogen; a fifth element, sulfur, is often present. Proteins, like starches and lipids, are built up from smaller units. The building blocks of proteins are *amino acids*; there can be several hundred amino acids in a single protein. Since there are 22 different amino acids, a tremendous variety of chemical structures among proteins exists. To get some idea of how many structures are possible, think of the number of ways in which the 26 letters of the alphabet can be arranged to form words!

Only 20 of the 22 amino acids are found in animals; the other two are limited to plants. Our bodies can manufacture half of the 20 amino acids from other food materials in our diet. However, our bodies cannot make the other ten but must obtain them from the proteins in the foods we eat. These ten are referred to as *essential amino acids* since they must be supplied in our diet. Without them, our cells could not manufacture the proteins needed for growth and repair, or for making enzymes and hormones. Consequently, our diets must include some protein.

The major source of protein in our diet is meat. But, what about vegetarians? Since most vegetables are low in essential amino acids, vegetarians must be very careful to select the right combination of plant foods that can furnish the requirements. By preparing meatless meals consisting of the proper mixture of beans, grains, seeds, and dairy products, vegetarians can obtain all the essential amino acids. Other foods that constitute a good source of essential amino acids are eggs, fish, cheese, and gelatin desserts.

How about testing some of your favorite foods for protein? The appearance of a purple or lavender color upon the addition of 3 ml of Biuret reagent to 2 ml or 2 gm of the food being tested indicates the presence of protein. If you use a solid sample, mix thoroughly after adding the Biuret reagent. You will have to be careful about interpreting your results when using food products with an intense color.

How to Soften a Tough Protein

Undoubtedly, you have eaten a tough piece of meat. If you knew that the meat was tough before you cooked it, you might have sprinkled some meat tenderizer on it. Ever wonder what effect the tenderizer has on the meat? Try the following experiment to see what the tenderizer does to proteins. Rather than experimenting with meat, where the effect would be more difficult to observe, you can use gelatin, which is mostly protein.

1. Following the directions given on the package, prepare some gelatin. Before the gelatin cools, carefully pour equal amounts into two small trays or baking pans.
2. After the gelatin has solidified, sprinkle meat tenderizer over one of the trays until there is a thin layer covering the gelatin.
3. Allow the two trays to remain undisturbed for five to ten minutes. Then gently poke the gelatin in each tray with a glass stirring rod. Explain your observations.

Try sprinkling the meat tenderizer on hot gelatin and then allow it to cool. What effect does the hot gelatin have on the meat tenderizer? Based on this observation, can you explain why knowledgeable chefs allow the meat and tenderizer to stand for several hours before broiling or barbecuing the meat?

Gelatin salads are prepared by mixing fresh fruits with gelatin. You might be interested in seeing what effect various fresh fruits, including pineapples, pears, peaches, and apples, have on gelatin. Simply place pieces of these fruits at different spots on a tray of gelatin after the gelatin has solidified. Why do cookbooks recommend against using fresh pineapple in gelatin desserts?

Let's Fry an Egg

Everyone knows what happens when you fry an egg. Obviously, the non-yolk portion turns white and becomes opaque as it is cooked. But can you explain what is happening in chemical terms?

To begin, start by testing the non-yolk part of an egg, called the albumin, to determine whether it is a sugar, starch, or protein. After determining what organic compound is present in the albumin, place an egg in a pan or pyrex bowl and fry it. After it cools, test the egg white to see if you get the same results.

What happened to the chemical composition of the albumin? Can you then explain why it changed in appearance after it was heated? Although the egg white retains its chemical composition, the various proteins present come together or *coagulate*. An interesting property of proteins is their ability to coagulate under certain conditions. Coagulation of proteins can be extremely important, as in the case of blood clot formation. If the proteins in blood did not coagulate, you could bleed to death from a simple cut!

The factors that cause proteins to coagulate and form a clot are quite complex and involve a series of chemical reactions. However, other factors promoting the coagulation of proteins, such as heat, can be examined in the laboratory or at home. Try adding different chemicals found in the kitchen to egg white to see which ones coagulate protein. Some you might try are salt, vinegar, sugar, ammonia, and alcohol. Which one, if any, is most effective in coagulating proteins?

Topics for Further Investigation

Each year, millions of people try to lose weight by going on a diet. Some diets recommend that you eat carbohydrate-rich foods, while others suggest meals consist of protein-rich foods. Books, magazine articles, and television shows feature new ways to lose weight quickly and permanently. Are any of these diets really effective? How is the body chemistry affected by each of these diets?

To lose weight, people often drink sugar-free sodas. Saccharin, a sugar substitute, is used. Only a small amount of saccharin is needed, since it is several hundred times sweeter than glucose. However, recent studies have raised concerns about the potential health hazard of saccharin. Although the findings are not conclusive, you may want to report on these research studies.

Many people have become cholesterol conscious. One factor associated with heart attacks is a high level of cholesterol in the blood. Examine the studies that suggest a relationship between heart attacks and high blood-cholesterol levels. Do scientists agree that high blood-cholesterol levels cause heart attacks?

Work with your school nurse or home economics teacher to plan the "perfect" three meals that would supply you with all the nutrients needed for good health, growth, energy, and repair of damaged or worn cells.

One of the greatest problems facing mankind is a potential food shortage. What are some of the possible solutions scientists are considering to avert the catastrophe of a worldwide famine?

Analyze the food served for lunch in the school cafeteria for a week. What are the principal organic compounds present? Do these meals supply you with the nutritional requirements? Can you make any suggestions to your school dietitian to improve their nutritional value?

*** Diabetes is a disease that prevents cells from getting enough glucose. Insulin, a protein hormone produced by the pancreas, is used to treat diabetes. Recombinant DNA technology has been used to obtain bacterial strains capable of producing insulin for treating diabetics. Research work is still needed to refine this process to obtain the most effective insulin. Much also remains to be accomplished in understanding the underlying causes of insulin-dependent diabetes and how insulin operates at the cellular level. After checking the scientific literature, design an experiment to provide additional information on diabetes.

*** Conduct a comparative study on the effects of diets high in saturated fats on the circulatory systems of various animals. What are effective antagonists to these saturated fats?

4 *Food Additives—Look Before You Eat!*

Pick up a loaf of bread, a box of doughnuts, or a package of breakfast sausages and check the label for the list of ingredients. Look for chemical compounds other than lipids, carbohydrates, or proteins. In examining the list of chemical ingredients on the labels, you may have discovered names of compounds such as sodium nitrate, calcium propionate, sodium benzoate, potassium sorbate, and butylated hydroxytoluene (BHT). These compounds are used to preserve foods so that undesirable chemical reactions do not occur. Without these preservatives, some of the chemicals in the food would change into new ones, which might cause serious illness if eaten. These chemical changes are brought about by small organisms, especially bacteria and yeast found in the air. Once in contact with food, these organisms cause chemical reactions. For example, bread without preservatives can spoil after several days, especially in hot and humid weather, as evidenced by a greenish mold formation caused by small organisms known as fungi. If bread is to be kept for some time, refrigeration would slow down the growth of the fungi responsible for the mold. To allow consumers to store bread for an even longer time, chemical preservatives are often added during the manufacturing process.

Methods to control the growth of small organisms that spoil food have been practiced for centuries. Salt and vinegar were the first chemicals used as preservatives, especially for keeping food during the winter season when hunting and farming were impos-

52

sible. These two chemicals inhibit the growth of small organisms. Drying was another early method used to preserve foods for long periods of time. Without water, bacteria and yeast cannot grow, reproduce, and thus contaminate the food. Early homemakers also smoked meats; this procedure coats the meat with chemicals which prevent the growth of small organisms. Smoking meat and fish is still popular today, not only to preserve them but also to give these foods a distinctive taste.

After the food was treated to destroy the bacteria and yeast, it was often placed in an environment where new organisms could not come into contact with the food. Placing the treated food in either sealed cans or jars was the most common method used to prevent it from becoming contaminated with small organisms. Since these containers were airtight, moisture was also kept out. Without water, new bacteria and yeast could not start to grow. Although salting, drying, smoking, and canning were effective methods for preserving foods, these procedures were time-consuming and often meant that people had to spend weeks preparing foods for storage. Today, many foods are first preserved by adding certain chemicals before they are packaged for sale.

The modern method of preserving food involves the addition of special chemical compounds to slow down the action of bacteria and yeast. Over the years, the list of chemical preservatives has grown to include a wide variety of compounds. Some of these chemicals used as preservatives are found naturally in our bodies and consequently do not pose a health hazard. However, some of those not normally present in our bodies have been shown to be potentially harmful if consumed in large quantities. When proven to be a possible health hazard to humans, a chemical preservative is either banned or restricted for special purposes by the Food and Drug Administration (FDA).

One chemical compound that can serve as a preservative is not only safe to consume but is also required for good health—vitamin C. Known chemically as ascorbic acid, vitamin C prevents fruit from turning brown upon exposure to the air. You have undoubtedly observed that apples, pears, bananas, and other fruits darken when they are peeled and left standing for a short time. Upon exposure to air, chemicals in the fruit react with oxygen, destroying cells and causing them to turn brown (see Figure 4–1). Vitamin C can slow down the reaction between the chemicals in the fruit and the oxygen in the air, thus preserving the color and taste of the fruit. Can you design an experiment to demonstrate the preservative action of vitamin C? Is vitamin C more effective with one particular fruit? Would fruit juice, which contains vitamin C, have

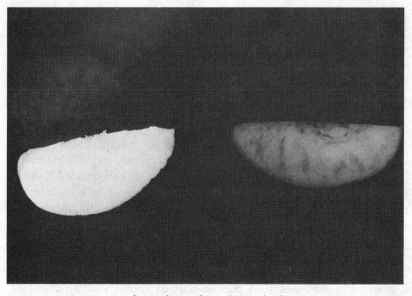

Figure 4–1. Can you explain why soaking the apple slice, pictured on the left, in a vitamin C solution prevented it from looking like the one on the right?

the same effect as vitamin C tablets? Why do chefs often squeeze a lemon or lime over a salad containing freshly cut apples?

BLT on White With Mayo—Hold the Preservatives!

Bread is one food that would quickly spoil without preservatives. Either calcium propionate or sodium propionate is usually added to the dough before baking to preserve the bread. The FDA has determined that the level of propionate in bread is sufficient to inhibit mold formation but low enough to be safe for human consumption. Can you design an experiment to test the effectiveness of the propionate preservative in bread? In planning your procedure, keep in mind that you will need to establish conditions to favor mold formation. Be sure to include breads that are prepared differently before making any conclusions about the effectiveness of the preservatives. How quickly will mold form on bread without preservatives?

How to Sniff Out a Preservative

You can check for the presence of either sodium propionate or calcium propionate in foods by using the following procedure. Since the amount of preservative is very small, it is difficult to check for the propionate compound directly. Consequently, to detect the preservative, you must first change it into another chemical compound which is easier to identify. This compound is called an *ester*. Esters are organic compounds that are easily identified by their pleasant odors. To make sure that you recognize the smell of esters, first use the following procedure with a piece of bread containing propionate preservative.

1. Break up the food into small pieces and place them in a beaker containing enough distilled water to cover the food. Keep the volume of water to a minimum.
2. While stirring, gently warm the food and water for five minutes. The warm water will extract any propionate preservative.
3. Carefully filter the solution through cheesecloth into another clean, dry beaker (see Figure 4–2). Place 10 ml of this liquid into a test tube and add 5 ml of 6M sulfuric acid. The letter *M* refers to the concentration or relative strength of the sulfuric acid. Since 6M sulfuric acid is concentrated, be *very* careful when pouring it into the test tube. If accidentally spilled, the acid can cause serious skin burns. Describe what happens after the acid is added.
4. Add 5 ml of ethyl alcohol to the test tube. Again be very careful, since the alcohol is flammable. Place a thermometer in the test tube and *gently* heat it. The alcohol will start to vaporize at approximately 80°C. As the temperature approaches 100°C, carefully wave some of the vapors toward your nose (see Figure 4–3). *Never* inhale any vapors by placing your nose directly over the test tube since the fumes may be quite powerful and pungent. Describe the odor.

Ester Smells Great!

The formation of an ester in the preceding experiment was used to identify a chemical preservative. However, esters are normally added as artificial flavorings to many foods. Candy and ice cream are two foods containing esters, which are produced by combining an alcohol with an organic acid. In the following exer-

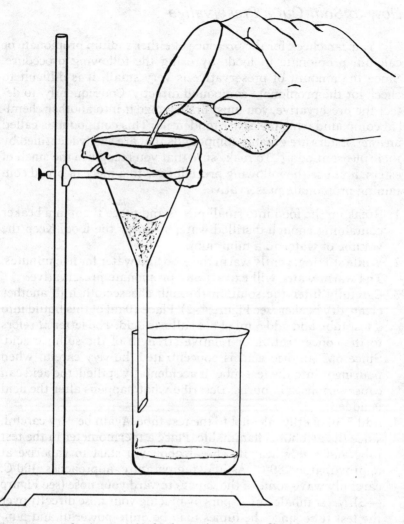

Figure 4–2. After heating the food sample in water, filter the mixture through cheesecloth. Any propionate preservative will be in the filtrate collecting in the beaker.

cise, you can try your hand at preparing some esters. Experiment to discover different smells (esters)! Place 2 ml of an alcohol (either amyl, methyl, ethyl, or isoamyl) and 2 ml of an organic acid (either acetic, salicylic, butyric, or decanoic) into a test tube. After adding 1 ml of 6M sulfuric acid, place the test tube in a boiling-water bath for a few minutes. Remove the tubes and carefully wave the vapors toward your nose. Try different combinations of an alcohol and an

Figure 4–3. When detecting odors produced by chemical reactions, never *inhale by placing your nose directly over the sample. Always wave some of the vapors toward your nose with your hand.*

organic acid to see what smell you can create. Keep a record of the alcohol and acid you mix, since you may produce an interesting odor! Some of the esters you produce may have a more distinctive odor than others. Can you identify any of these? You may have to work quickly in some cases to detect the ester since the odor may be either faint or short-lived. What industry, other than food processing, relies upon esters?

Not Everything Smells Like Ester!

Dried fruits would spoil very quickly without preservatives. To prevent the growth of bacteria and mold on dried apricots, pears, peaches, and apples, food processors add sulfur dioxide as a preservative. Sulfur dioxide, like vitamin C, prevents the formation of the brown color produced when the chemicals in the fruit spoil by reacting with oxygen in the air. However, when present in high concentrations, sulfur dioxide is harmful, has an unpleasant odor, and is considered an air pollutant. Naturally, the FDA requires that only small amounts of sulfur dioxide can be added to preserve foods. In the following exercise, you can check dried fruits, and any other foods you wish to test, for the presence of sulfur dioxide.

1. Place several pieces of dried fruit in a large beaker and cover them with distilled water. Allow the fruit to soak overnight.

2. The next day, filter the liquid and fruit through a piece of cheesecloth and collect the filtrate in a clean, dry flask. Using a glass stirring rod, squeeze the fruit to extract as much juice as possible.

3. Determine the volume of the filtrate and add an equal volume of 3 percent hydrogen peroxide to the filtrate. After stirring for a few minutes, add a saturated solution of barium chloride drop by drop. If sulfur dioxide is present as a preservative in the fruit, solid material, known as a *precipitate*, will form. This precipitate is produced by a chemical reaction between the preservative and the barium chloride. Continue adding the barium chloride solution until no more precipitate forms. Since the amount of sulfur dioxide added as a preservative to foods is very small, you may have to allow the contents of the flask to remain undisturbed overnight. The next day, examine the bottom of the flask for the presence of a white precipitate. How would you modify this procedure if you wanted to determine the percentage of sulfur dioxide preservative in the food sample? Find out how sulfur dioxide acts as an air pollutant when present in high concentrations.

Food Additives That Fortify

Chemicals are added to foods not only as preservatives but also as fortifiers to supply your body with important chemicals. As you may recall from your study of foods, the major nutrients required for good health and growth are lipids, carbohydrates, and proteins. However, your body also needs small amounts of other types of chemicals; these include vitamins and minerals. In some cases, manufacturers add these chemicals to foods as supplements. Have you ever heard of "enriched" bread or "fortified" cereals? For example, a manufacturer may add iron to "enrich" bread or vitamins to "fortify" cereals. Check the labels on various food packages to see if any vitamin or mineral supplements are present. What foods usually contain supplements that "enrich" or "fortify"?

Iron is one chemical often used as a "fortifier" in foods. Usually added to breakfast cereals and bread, iron is needed to make hemoglobin. Hemoglobin, which is found in red blood cells, transports oxygen to the cells where it is required for energy production. Without sufficient iron, your red blood cells could not supply the cells with enough oxygen. The result would be a disease known as anemia, which is characterized by a lack of energy. Only a small amount of iron is required in the diet to enable the hemoglobin to supply the cells with sufficient oxygen. Consequently, well-balanced meals provide enough iron for most everyone's needs. However, many people may periodically require additional iron, obtained either by eating foods such as liver, spinach, or raisins, or by taking iron supplement pills.

Iron It Out Yourself

In the following experiment, you can test for the presence of iron in foods. If iron is present, it will react with the chemical compound potassium thiocyanate to form a red color. The more iron present, the darker the color. Before testing various foods, you may want to check out this procedure with vitamin tablets containing iron. Soak the tablets in water to dissolve any colored coating. Allow to dry, then crush two tablets into a powder, using a mortar and pestle as illustrated in Figure 4–4. If you plan to compare the iron content of vitamin pills with that of different foods, do not discard any of the red-colored solutions until you are completely finished with this analysis.

1. Place 3 gm of a pulverized food sample in a crucible and heat strongly until only an ash residue remains. After the crucible has cooled, scoop the ash residue into a beaker, add 10 ml of distilled water, and stir for several minutes.

2. Filter the solution and collect the filtrate in a small beaker or large test tube. Add a few drops of concentrated nitric acid

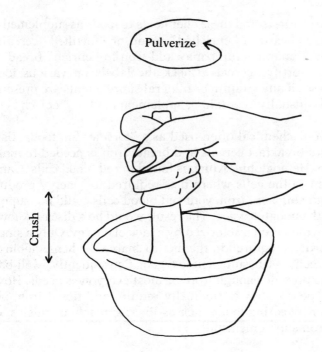

Figure 4–4. Using the pestle in a circular motion will pulverize the sample in the mortar. Moving the pestle up and down will crush any large substance before pulverizing it.

followed by 2 ml of 0.1M potassium thiocyanate solution to the filtrate and observe whether a red color is produced. How does the color formed from testing the vitamin pills compare with that produced in testing the food sample?

Can you determine how much more iron is present in the vitamin pills than in the food sample? What should be added to the colored solution containing iron from the vitamin pills to make it the same shade of red as the one produced from testing the food sample? To compare the concentration of iron between all the samples tested, you must record the volume added to dilute the red color for each sample. For example, if you add 100 ml to the solution containing iron from the vitamin pills to get the same shade of red as present in the solution from the food sample, then the vitamin pill has 100 times more iron than the food sample. If this turns out to be the case, remember that you are using a small amount of food for the iron analysis. Check a reference source to see how much iron is supplied by an average-sized serving of the foods you tested.

Vitamins—Alphabet Power

If you eat well-balanced meals, you should get all the vitamin A, B, C, D, E, and K your body needs. To accomplish this, your daily meals should include foods from each of the four basic categories: meat, poultry or fish; fruits or vegetables; grains or cereals; and dairy products. The proper combination and balance of these four classes of food should provide all the vitamins your body requires for good health. But do your meals consistently include foods from each of these categories? Obviously, most of us don't always take the time to eat foods that would supply enough vitamins. In addition, we sometimes rely on "fast foods" that provide little, if any, nutritional value. Moreover, modern methods of food processing destroy some of the vitamins present in food. Frozen, canned, and precooked foods contain drastically reduced levels of vitamins.

Since there are times when we fail to get enough vitamins from our meals, we may resort to taking a multi-vitamin pill. Or, if we are aware that the foods we have eaten are not providing enough vitamins, we might make a point of eating something that has been fortified with vitamins. After scientists recognized the importance of vitamins to good health, food manufacturers began to add them as supplements to various foods. Scientists also discovered that well-balanced meals might not provide sufficient vitamins under certain circumstances. For example, small children and those adults with limited exposure to sunlight were more likely to suffer from a disease caused by a lack of vitamin D. Cholesterol, a lipid present in nerves and glands, is converted to vitamin D when the skin is exposed to the ultraviolet rays from the sun (see Figure 4–5).

Insufficient levels of vitamin D can cause rickets, a disease characterized by abnormal bone development and a slower rate of growth in children. Without sufficient vitamin D, bones are softer and more likely to break from a simple fall. Since the most important source of vitamin D from foods is fish, especially the liver oils from cod and halibut, many people would not receive the required amount of vitamin D in their diets. Besides depending upon the conversion of cholesterol through the action of sunlight, we rely upon the vitamin D added as a food supplement, especially in milk. Food processors also add vitamin D to infant formulas, breakfast cereals, margarine, breads, and flour. Consequently, most people receive their required vitamin D by the action of sunlight on cholesterol present in skin cells and as a food supplement.

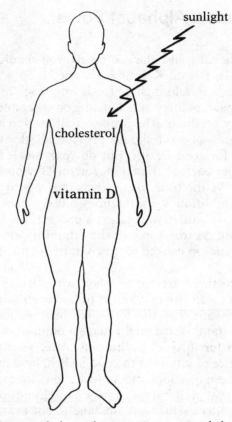

Figure 4–5. Sunlight converts cholesterol, present in nerves and glands in the skin, into vitamin D.

With a precipitous decline in the number of people with vitamin D deficiencies, rickets has practically been eliminated except in areas of the world where malnutrition exists.

In addition to vitamin D, we require several other vitamins in order to prevent deficiency diseases. In some cases, our bodies can manufacture some of the vitamins, but often in amounts too small to be helpful. But since our bodies cannot synthesize all the vitamins necessary for good health, we must rely upon a continuous supply in our diets. However, we are not the only ones who would pay the consequences of a diet low in vitamins. In fact, animals other than humans can become sick and even die without enough vitamins. Although the exact function of all the vitamins is not known, scientists have discovered the relationship between vitamins and certain biological processes. Some of the more common vitamins, along with their function, foods in which they are concentrated, and their relationship to deficiency diseases are listed in the following.

VITAMINS AND CHEMICAL REACTIONS IN THE BODY
Vitamins affect a wide variety of chemical reactions necessary for good health.

- **Vitamin A** aids in the growth and repair of skin cells; essential for enabling the eyes to adjust to different light intensities. Food sources: whole milk, butter, egg yolks, cheese, yellow and green vegetables (especially carrots). Deficiency diseases: rough skin, night blindness
- **Vitamin B** involved in chemical reactions producing energy; needed for normal oxygen transport by hemoglobin in red blood cells. Food sources: whole grains, dark green leafy vegetables, meat. Deficiency diseases: muscle weakness, skin problems (especially of the mouth and tongue), anemia
- **Vitamin C** required for healthy bones and teeth; role in preventing colds is disputed. Food source: citrus fruits, tomatoes, and potatoes. Deficiency disease: scurvy (gum disease)
- **Vitamin D** needed for normal bone development. Food sources: whole milk, fish-liver oils. Deficiency disease: rickets
- **Vitamin E** aids in maintaining healthy red blood cells; function in slowing down the aging process is questioned. Food sources: vegetable oils, whole grains, green vegetables, meat. Deficiency disease: decrease in red blood cells
- **Vitamin K** involved in blood clot formation. Food sources: green vegetables, cauliflower. Deficiency disease: excessive bleeding from simple cuts

Let's Look at Labels

By law, any food product advertised as being "enriched" or "fortified"—with vitamins, for example—must list all the ingredients on the package label. If you look at the label, you'll notice quite a bit of nutritional information that can be helpful to the consumer. Examine the label on a loaf of enriched sliced white bread. How many slices constitute a serving? What organic nutrients are present and how many grams of each are supplied per serving? Check the label for the term "percentage of U.S. recommended daily (or dietary) allowance." This phrase, abbreviated U.S. RDA, signifies the official government estimate of the amount of nutrient each person needs to consume daily in order to remain healthy. The U.S. RDA is not intended to be a guideline outlining the absolute minimum amounts of nutrients that everyone must have in their diets. Rather, the U.S. RDA is a recommendation about how much of any given nutrient should be present for the "average" person. Since the nutritional requirements of an individual will vary from time to time, depending on age, weight, health status, and medical history, the U.S. RDA figures should be taken as recommended guidelines and not as absolute requirements. With this in mind, take another look at the label on the loaf of bread.

How many grams of protein are supplied per serving? What percentage of the U.S. RDA does this amount constitute? Can you calculate how many total grams of protein are recommended on a daily basis? How many slices of bread would you need to eat in order to supply all your required protein? Check the labels on different brands of enriched sliced white breads to compare the vitamins they provide as additives. In conducting your analyses, be aware that thiamine, riboflavin, and niacin are different types of vitamin B. Is one brand closer to the U.S. RDA for any one vitamin? Does one supply a greater U.S. RDA of all the vitamins? You may also find it interesting to conduct the same analysis with different brands of breakfast cereals. With either breads or cereals, you may find it helpful to make up a table, indicating the brand name, vitamins added, the percent of U.S. RDA for each vitamin supplied per serving, and cost per serving. Based on information listed in the table, would you now make it a point to purchase a particular brand of bread or cereal?

Nutrients and Vitamins

The following program will help you determine the additional amount of organic nutrient or vitamin needed after eating one serving of any bread or cereal. For example, if you eat two slices of bread supplying 4 gm of protein, which is 6 percent of the U.S. RDA, this program would reveal that you need an additional 63 gm of protein to meet the U.S. RDA. If you are analyzing vitamin requirements, type in the word "UNITS" in place of "GRAMS" in lines 50, 90, 160, 190 and 200. If you have run any of the programs in the preceding chapters, you will notice something slightly different in this one. The program has been written so that you will have only two chances to supply the correct answer. If you fail to do so, the computer will show you the correct answer and how to arrive at it. By the way, round off any decimal answer to the lower whole number and do not enter zero for any value.

If you have saved any previous programs, you might like to try to modify them so that only two chances are given to arrive at the correct answer. Can you figure out which lines in the following program must be added to previous ones and where they must be placed? Use commands to clear the screen after entering correct and incorrect answers.

```
10 PRINT "WHAT IS YOUR FIRST NAME";:INPUT NA$
20 PRINT
30 PRINT "TYPE IN THE NUTRIENT OR VITAMIN YOU"
40 PRINT "ARE ANALYZING";:INPUT C$
50 PRINT "HOW MANY GRAMS OF ";C$
60 PRINT "ARE SUPPLIED PER SERVING";:INPUT GM
70 PRINT "WHAT PERCENT OF THE U.S. RDA IS"
80 PRINT "SUPPLIED PER SERVING";:INPUT PE:PE =
   PE/100:PRINT
90 PRINT "HOW MANY ADDITIONAL GRAMS OF ";C$
100 PRINT "DO YOU REQUIRE";:INPUT RD
110 IF RD = INT ((GM/PE) − GM) THEN 250
120 IF X = 1 THEN 150
130 PRINT "SORRY";NA$;", THAT IS NOT CORRECT. TRY
   AGAIN."
140 X = X + 1:GOTO 90
150 PRINT "THAT IS ALSO INCORRECT, ";NA$;"."
160 PRINT "THE CORRECT ANSWER IS ";INT((GM/PE) − GM);
   "GRAMS."
170 PRINT "REMEMBER THAT THE TOTAL AMOUNT REQUIRED"
180 PRINT "IS CALCULATED BY DIVIDING THE PERCENT"
190 PRINT "OF U.S. RDA INTO THE GRAMS PER SERVING."
200 PRINT "YOU THEN SUBTRACT THE GRAMS SUPPLIED"
210 PRINT "FROM THIS NUMBER TO GET THE ADDITIONAL"
220 PRINT "AMOUNT REQUIRED.":PRINT
230 GOTO 260
240 PRINT
250 PRINT NA$;", THAT IS CORRECT!"
260 PRINT "WOULD YOU LIKE TO TRY ANOTHER PROBLEM (Y OR
   N)";:INPUT AN$
270 IF LEFT$(AN$,1) = "Y" THEN X = 0: GOTO 20
280 END
```

Add Color to Your Life, But Not to Your Food!

Reds, greens, blues, and yellows are attractive colors that, when added to food products, make them look more appetizing. Consequently, food processors rely upon coloring agents as food additives in an attempt to catch the consumers' attention. Many of these dyes are prepared by extracting chemicals from coal tars and converting them into synthetic compounds that can be used as colorful additives. However, research studies have indicated that some of these chemicals can cause cancer in animals. As a result, the FDA has ordered that their use as coloring additives be stopped; this was the case in 1976 when red dye number 2, a commonly used coloring additive, was banned by the FDA.

To avoid synthetic coloring agents, many consumers buy foods containing only natural dyes. Natural food colors are present in beets, red cabbage, carrots, lemons, oranges, and, of course, green vegetables. Can you decorate a cake with icings prepared with only natural dyes? How colorful can you make it? Experiment with different procedures to extract the dyes from colorful foods. You may first want to chop or grind the food to make it easier to extract the natural dyes by boiling the food in water or alcohol. If you use alcohol, allow it to evaporate, leaving a dye that can then be dissolved in water. Try mixing different dyes to see what colors you can produce. How many different colors does your cake have? Good eating, and remember, it's healthy because there are no synthetic color additives—at least not in the icing!

Topics for Further Investigation

Additives are occasionally used to enhance the flavor of foods. Monosodium glutamate (MSG) is probably the most widely used flavor enhancer. Why has the FDA required that MSG be listed on the label and banned its use in baby foods?

Preserve some foods without the use of chemicals. Be sure to include foods from the four basic categories so that all the required nutrients will be supplied on a daily basis.

Examine how the FDA investigates the potential health hazard of a food additive. What experimental procedure is followed before the FDA states that a particular chemical is carcinogenic (cancer-causing)?

Keep track of what you eat in one day. List all the chemical additives you have consumed in these foods. What is the purpose of each additive?

Visit a "health food" store. Can you find any foods with additives that either preserve, enrich, fortify, or enhance? Talk to the manager to find out whether any special procedures must be followed in marketing "health foods." Do they have a shorter shelf-life than foods in a supermarket?

Although only small amounts of vitamins are required for good health, some people take mega-vitamin tablets that far exceed the U.S. RDA recommendations. Research studies have shown that excessive vitamin intake can cause health problems. Examine the recent reports and research literature to discover which vitamins have been shown to be a health hazard when taken in large doses.

*** Many vitamins function as coenzymes in biochemical reactions. Although scientists have been investigating the mechanism by which enzymes and their coenzymes operate, many details of their biochemical action remain to be understood. Select a vitamin that serves as a coenzyme and design a research project aimed at elucidating its role in all the biochemical pathways in which the vitamin is involved.

*** Can you design an experimental system to test the effects of either chemical preservatives or large doses of specific vitamins on metabolism or developmental patterns?

CHEMISTRY IN THE BATHROOM

5 *The Aspirin Family*

A quick look inside your medicine cabinet would reveal numerous chemical compounds used almost every day by someone in your family. These might include drugs prescribed by a doctor for curing an infection, treating an eye irritation, or relieving the discomforts caused by allergies. In addition to prescription medications, you would also find a variety of over-the-counter drugs, including pain relievers, stomach-acid neutralizers, and vitamin supplements. With so many different medications available at one time or another in most homes, an understanding of their chemical nature and how they function in the body is important. In view of the increasing use of nonprescription drugs, such knowledge can be helpful to consumers, especially in making purchases based on practical information rather than advertising claims.

The nonprescription drug taken most often is aspirin. In fact, aspirin is the most widely used drug in the world. Unlike most medications, which are limited to the treatment of one specific disease or ailment, aspirin is used for a variety of medical problems. Aspirin reduces fever, pain, and the swelling of tissues caused by arthritis. Research studies have shown that aspirin can also be effective in preventing both blood clots and the recurrence of kidney stones. No other drug covers such a broad spectrum of diseases or ailments.

Long before its actual chemical nature was known, people relied upon the wide-ranging power of aspirin. The first use of aspirin can be traced to the early Romans and Greeks, who discovered that the bark, fruit, and leaves from certain shrubs and trees were beneficial in treating a variety of ailments. They prepared extracts from different plants, especially the willow tree, to treat earaches, battle wounds, and eye diseases. The use of plant extracts continued for centuries in countries throughout the world. American Indian tribes, for example, found a liquid extract prepared from willow bark helpful in reducing fever and alleviating pain. Even though such extracts were used for hundreds of years, the chemical nature of the active ingredient in medicinal plants remained unknown.

The chemical composition of the ingredient in these plants was discovered in the nineteenth century when chemistry was born as a science. Chemists succeeded in isolating and identifying the ingredient; they named it salicylic acid. Once its identity had been revealed, pure salicylic acid was available for general use. Unfortunately, the high cost of isolating and purifying salicylic acid from plants prevented it from becoming more widely used as a pain reliever. Moreover, salicylic acid was found to be extremely irritating to the lining of the stomach. Those who took it to relieve pain often suffered stomach irritation or nausea and occasionally developed ulcers. Chemists immediately began to search for a way to eliminate the side effects of salicylic acid without reducing its medicinal value.

In 1853, a chemical reaction capable of changing salicylic acid into another compound was discovered. However, the significance of this process was not immediately recognized. Approximately 40 years later, a young scientist named Felix Hoffman was searching for a pain reliever for his father, who was suffering from crippling arthritis. Following up the earlier reports describing the chemical process that changed salicylic acid into a different compound, Hoffman conducted experiments both in his laboratory and on his father. Satisfied that he had an effective pain reliever that did not cause any serious side effects, he approached a chemical company with his findings. Called acetylsalicylic acid, this new pain reliever was introduced to the public at a reasonable cost by the Bayer Company in 1899 under the name Aspirin.

To make aspirin, salicylic acid is reacted with acetic acid. If aspirin tablets are kept for several months, the opposite or reverse reaction occurs. The acetylsalicylic acid reacts with water in moist air to break down into salicylic acid and acetic acid. When this reaction occurs, aspirin tablets smell like vinegar, which is dilute acetic acid. Since the acetic acid is likely to irritate the lining of your stomach, any aspirin tablets that smell like vinegar should be discarded. Wise consumers always purchase small bottles of aspirin to avoid keeping them in the medicine cabinet for too long. Another good idea is to keep the bottle tightly capped to prevent the moisture in air from reacting with the acetylsalicylic acid to form acetic acid.

Aspirin does not have to change into acetic acid in order to be a stomach irritant for some people. The irritation, caused by the acids in the aspirin tablets, can be avoided by taking the aspirin with a large glass of water or milk. The water dilutes the acids so that their strength, or concentration, is decreased. Milk coats the stomach lining to reduce irritation caused by the acids.

How Weak or Strong? Only the pH Can Tell

The strength of an acid is measured by units that make up the *pH scale*. Scientists use the pH scale to indicate whether an acid is strong or weak. In addition, the pH scale is also related to the strength of another group of chemical compounds known as bases. The pH scale ranges from 0 to 14, as shown in Figure 5–1. Any solution whose pH falls right in the middle of the scale, with a pH value of 7, is neutral. The acid range is below 7; the strongest acids have the lowest pH values, while the weakest ones are closer to 7.

Any solution with a pH value above 7 is a base; the strongest bases have the highest pH values, while the weak bases are closer to 7. Each value on the pH scale represents a tenfold difference in relative strength. For example, an acid with a pH value of 4 is ten times stronger than one with a pH value of 5. What is the differ-

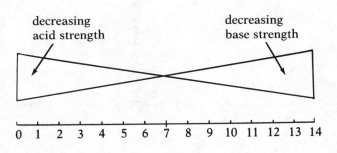

Figure 5–1. The pH scale, ranging from 0 to 14, indicates the concentrations of acids and bases. Each unit on the pH scale represents a tenfold difference in relative strength.

ence in relative strength between a basic solution with a pH value of 8 and one with a pH value of 6?

To check your understanding of the pH scale, enter the following program into the computer. If your computer accepts both upper and lower case letters, type "pH." After providing the initial pH value, you are asked whether you wish to increase or decrease the acidity. You can change the acid strength by a factor of 10, 100, or 1,000. Can you then figure out what the new pH value will be? Think before supplying your answer since you will have only two chances to get the correct pH value. Use the computer command appropriate for your system if you wish to clear the screen after each problem or run.

pH

```
10 PRINT "WHAT IS YOUR FIRST NAME";:INPUT NA$
20 PRINT "WHAT IS THE INITIAL PH VALUE";:INPUT N
30 PRINT "DO YOU WISH TO INCREASE OR"
40 PRINT "DECREASE THE ACIDITY (I OR D)";:INPUT AN$
50 IF AN$ = "I" THEN GOTO 210
60 PRINT "DO YOU WISH TO DECREASE THE ACIDITY"
70 PRINT "BY A FACTOR OF 10, 100 OR 1000";:INPUT D
80 IF D = 10 THEN N = N + 1
90 IF D = 100 THEN N = N + 2
100 IF D = 1000 THEN N = N + 3
110 PRINT "WHAT IS THE NEW PH VALUE";:INPUT P
120 IF P = N THEN GOTO 360
130 X = X + 1
140 IF X = 2 THEN GOTO 180
150 PRINT "SORRY THAT IS NOT THE CORRECT ANSWER."
160 PRINT "TRY AGAIN, ";NA$;".":PRINT
170 GOTO 110
180 PRINT "THAT IS ALSO INCORRECT, ";NA$;"."
190 PRINT "THE CORRECT ANSWER IS";N
200 GOTO 370
210 PRINT "DO YOU WISH TO INCREASE THE ACIDITY"
220 PRINT "BY A FACTOR OF 10, 100 OR 1000";:INPUT I
230 IF I = 10 THEN N = N - 1
240 IF I = 100 THEN N = N - 2
250 IF I = 1000 THEN N = N - 3
260 PRINT "WHAT IS THE NEW PH VALUE";:INPUT A
270 IF A = N THEN GOTO 360
280 X = X + 1
290 IF X = 2 THEN GOTO 330
300 PRINT "SORRY, THAT IS NOT THE CORRECT ANSWER."
310 PRINT "TRY AGAIN, ";NA$;".":PRINT
320 GOTO 260
330 PRINT "THAT IS ALSO INCORRECT, ";NA$;"."
340 PRINT "THE CORRECT ANSWER IS ";N
350 GOTO 370
360 PRINT:PRINT "VERY GOOD ";NA$;", YOU ARE CORRECT."
370 PRINT "DO YOU WANT TO TRY ANOTHER PROBLEM";:
    INPUT PR$
380 X = 0
390 PRINT
400 IF LEFT$(PR$,1) = "Y" THEN GOTO 20
410 END
```

An Indication of How You're Doing

Indicator solutions or papers are used to determine the pH value of solutions. Obtain some universal pH paper and measure the pH values of various solutions found in the home by dipping a small piece of pH paper into the solution and then comparing it with the test samples of the vial (see Figure 5–2). You may want to check milk, sodas, vinegar, ammonia, coffee, milk of magnesia, and tomato juice. Also check the pH value of both tap and distilled water. Can you explain the reason for the difference in pH values for these two types of water? Crush some aspirin tablets in water and check the pH of the solution. If you can obtain any aspirin tablets that smell like vinegar, you may be interested in comparing their acidity to tablets that have not changed into acetic acid. How would you prove that each unit on the pH scale represents a tenfold change in relative strength? Using an acid and a base, can you make a neutral solution with a pH value of 7?

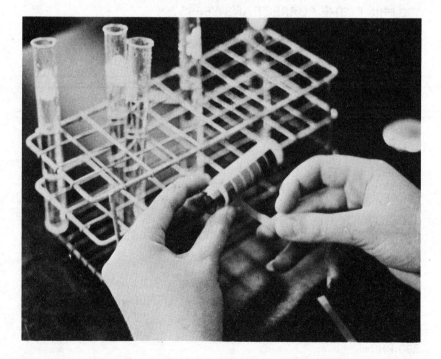

Figure 5–2. A piece of universal indicator paper is compared with samples on the vial to determine the pH of the solution.

Red Cabbage—A Natural Indicator

Universal pH paper is prepared by adding a dye that changes color at different pH values. Dyes can be extracted from a variety of foods, including red cabbage, beets, grapes, and blueberries. Purple and blue flowers can also be used as sources for dyes. In the following experiment, you will extract the dye from red cabbage to use as an indicator with various solutions found in the home.

1. Chop the leaves from a red cabbage into small pieces. Place the pieces into a blender and cover with water. Continue blending until the contents have liquefied.
2. If a blender is not available, place the small pieces of red cabbage in a beaker, cover with water, and slowly boil the contents until the water turns a dark color. You may have to add some water to prevent the cabbage from burning. After allowing the beaker to cool, carefully strain the mixture through cheesecloth to collect the liquid in a clean, dry beaker.
3. Mix 3 ml of the colored solution in a test tube with 3 ml of white vinegar. What color does the cabbage solution turn? Check to see what color changes are brought about by adding various other household solutions, including ammonia, lemon juice, and seltzer water or club soda. Based on the color changes, which solutions can be grouped together as having similar chemical properties?

A Recipe to Relieve Headaches

In Chapter 4, you may have prepared different esters by reacting an organic acid with an alcohol in the presence of sulfuric acid. Aspirin can also be prepared by a similar process when acetic acid combines with the alcohol part of salicylic acid. However, in place of acetic acid, another compound, acetic anhydride, is used to make the reaction proceed more quickly. The product will be acetylsalicylic acid, or simply aspirin.

1. Place 5 gm of salicylic acid in a flask. Carefully add 20 ml of acetic anhydride or 80 percent acetic acid. Then add three drops of 6M sulfuric acid to the flask. You may recall that the letter M refers to the relative strength of the acid. Since 6M sulfuric acid is strong, be careful when adding the acid since it can cause serious skin burns! Stir the contents of the flask for several minutes.

2. Swirl the contents of the flask. Insert a thermometer into the flask and heat the contents for ten minutes at 85°C (see Figure 5–3). After allowing the flask to cool, place it in an ice bath and add 100 ml of distilled water. Record your observations. Collect the precipitate on filter paper.

3. Allow the precipitate to dry overnight. To determine if all the salicylic acid has reacted, dissolve a small amount of the pow-

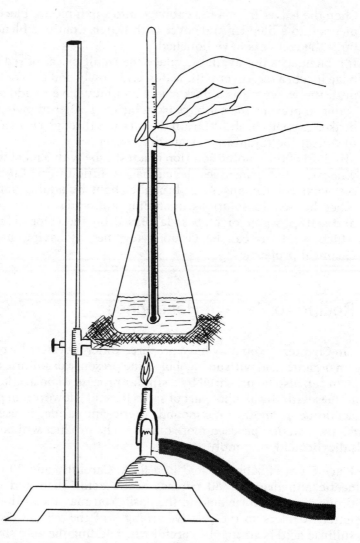

Figure 5–3. Heating for ten minutes at 85°C in acetic and sulfuric acids will dissolve the salicylic acid.

der in distilled water and add three drops of 1 percent ferric chloride or 1 percent ferric nitrate solution. A pink, light purple, or green color indicates the presence of salicylic acid. Does your aspirin contain any salicylic acid or has it all been changed into acetylsalicylic acid?

Aspirin—Which Brand Is Most Effective?

Is one brand of aspirin more effective in relieving pain and reducing fever? A chemical investigation will give you the answer! The following analysis of commercial aspirin tablets will explore their chemical composition. The information obtained from this exercise might be useful when purchasing your next bottle of aspirin.

1. As previously mentioned, the active ingredient in aspirin is acetylsalicylic acid. If you examine the label on the package or bottle, you will notice that the manufacturer lists the number of grains of aspirin per tablet. One grain is equivalent to 64.8 mg or 0.0648 gm of aspirin. Most tablets contain five grains of aspirin. How many grains of aspirin per tablet are present in various brands? Be sure to include nationally advertised, supermarket, buffered, and childrens' brands in your analysis. Calculate the grams of aspirin (acetylsalicylic acid) present in each tablet by multiplying the number of grains by 0.0648. As accurately as possible, weigh one tablet of each brand. Calculate the percent of aspirin in each tablet by using the following equation:

$$\frac{\text{weight of aspirin (gm)}}{\text{weight of tablet (gm)}} \times 100 = \text{percentage of aspirin per tablet}$$

Which brand contains the highest percentage of aspirin? Which brand has the lowest? Can you suggest a reason for this? You may find it helpful to construct a table, indicating the brand name, amount of aspirin per tablet, weight of a single tablet, and the percent of aspirin in a tablet.

2. As you discovered in the preceding analysis, a tablet is not 100 percent aspirin. Manufacturers use "fillers" to give the tablet some bulk and to prevent it from crumbling. Both sugars and starch may be used as fillers. In the next part of this analysis, you will test each brand for the presence of sugars and starch. Crush two tablets of each brand and add them to 15 ml of

distilled water. Before checking for fillers, test each aspirin sample with a small piece of universal pH paper. Since aspirin will not dissolve in water, make sure that you shake the test tubes to disperse the powdered aspirin before dipping the pH paper in the water. Which brand has the lowest pH (is the most acidic)? Which brand is closest to neutral or pH 7? Can you explain what "buffered" aspirin is? Who should take buffered aspirin? After checking the pH values, you are ready to test for fillers.

3. Equally divide the 15 ml of the distilled water containing aspirin into three test tubes. Add five drops of Lugol's iodine to one test tube. A blue-black color indicates the presence of starch.

4. To the second test tube, add saturated sodium carbonate solution a drop at a time, testing with blue litmus paper after adding each drop. Stop adding the sodium carbonate solution when the blue litmus paper no longer turns pink. The litmus paper is an indicator; blue litmus paper turns pink in acidic solutions but remains blue in basic solutions. Once the acidity of the aspirin has been neutralized by the sodium carbonate, you can test for the presence of sugars. Add 5 ml of Benedict's solution to the neutralized aspirin solution and heat in a boiling-water bath for five minutes. Sugar is present if the solution turns yellow, orange, green, or red. With quantitative Benedict's solution, the darker the color, the more sugar present. Which brands contain fillers?

5. To the remaining test tube, add ten drops of 1 percent ferric chloride or 1 percent ferric nitrate. A pink, light purple, or green color indicates the presence of salicylic acid. Do any of the brands contain salicylic acid, which is irritating to the stomach? Make a table indicating the brand name, the pH value, and whether sugar, starch, or salicylic acid is present.

6. Finally, calculate the cost per tablet for the various brands. Based on the results of your analyses, is there one brand of aspirin you would make a point to purchase?

Extra-Strength Tablets—Super Power for Fast Relief

Undoubtedly, you've had a headache where the normal adult dosage of two regular aspirin tablets didn't provide much relief. After waiting several hours, you may have decided to try a brand advertised as having extra strength or one promising fast pain relief. Do these brands have a special chemical ingredient, other than acetylsalicylic acid, to reduce pain? If you check the label for a list of ingredients, you'll discover that these tablets contain the same active ingredient found in regular brands. The only difference is the presence of more grains of aspirin in an extra-strength tablet. But is it worth paying more money for extra-strength tablets?

To find out, compare the cost per tablet for regular and extra-strength brands. Then, divide the cost per tablet by the number of grains of aspirin in each tablet to determine how much you are paying for each grain of acetylsalicylic acid. Would it be advisable for the person pictured in Figure 5–4 to take three regular tablets rather than two extra-strength aspirins? You would have to compare the grams of aspirin provided in each case and also consider the cost.

Conduct a chemical investigation of various extra-strength brands similar to the analysis conducted on regular aspirin tablets. How do their pH values compare? Do any contain sugar or

Figure 5–4. Do three regular aspirin tablets contain the same amount of various chemical ingredients as two extra-strength tablets? The student is conducting an analysis to determine the answer.

starch as fillers? If any extra-strength brand is advertised as also being fast-acting, you can check this claim by comparing the time required for two tablets to dissolve in 1N hydrochloric acid (see Figure 5–5). The letter N refers to the relative strength of the acid; 1N hydrochloric acid has a pH value comparable to the acidity level of the chemical compounds involved in digestion in your stomach. Consequently, the tablets that dissolve the quickest would be the first to enter the bloodstream and be transported to the brain to relieve pain.

Chemical Pinch Hitters for Aspirin

Some people are allergic to aspirin and can suffer serious side effects, including stomach bleeding, from taking two regular tablets. Those with ulcers must also avoid aspirin since it would easily irritate their stomach lining. To relieve pain and reduce fever, these people must rely on aspirin substitutes. Various aspirin-free products are available. At first, manufacturers substituted a chemical compound called phenacetin for aspirin. Although it is as effective as aspirin, phenacetin was found to cause kidney damage if taken for long periods of time. Any aspirin substitute containing phenacetin must carry a warning on the label indicating the potential health hazard posed by using it for any prolonged period of time.

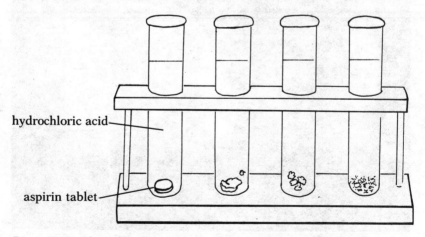

Figure 5–5. When placed in 1N hydrochloric acid to simulate conditions in the stomach, an aspirin tablet will slowly disintegrate. The faster the tablet goes into solution, the sooner it's absorbed into the bloodstream to provide relief.

Searching for a better substitute, many drug companies have switched to a different chemical, acetaminophen. Acetylsalicylic acid, phenacetin, and acetaminophen have similar chemical structures. Check a chemistry reference book for the structural formulas for each of these compounds. What similarities do you notice? How do they differ from one another?

Aspirin Relatives

Salicylic acid, the chemical originally used as a pain reliever, is the basis for making methyl salicylate, a compound found in liniments used to relieve muscular pain and aches. This medication works by enlarging the blood vessels and producing a feeling of warmth in the muscles. To produce methyl salicylate, methyl alcohol is combined with salicylic acid. If you check the procedure for making aspirin, you'll notice that the preparation for both acetylsalicylic acid and methyl salicylate involves the formation of esters. Following the procedure for producing an ester (page 56), react methyl alcohol with salicylic acid. Can you identify the odor produced by the ester, methyl salicylate?

Topics for Further Investigation

Although aspirin has been used to relieve pain for centuries, scientists are still uncertain about how aspirin works in the body. Check recent studies that have investigated the chemical process by which aspirin is believed to operate in reducing pain.

Report on the history of medicinal plants. Include the plants used most often by past civilizations to treat both serious diseases and minor ailments. Have these plants served as the original source for any modern drug apart from aspirin?

Investigate other medicinal properties of plants, besides pain relief. For example, you can test for the presence of antibiotics and determine how effective they are in preventing the growth of bacteria.

Since billions of aspirin tablets are manufactured daily, chemists needed a source of salicylic acid other than willow bark. To-day, aspirin is produced from petroleum extracts. Check to find out what other health-related chemical products depend upon petro-leum as a starting point. Contact a drug manufacturer to discover the chemical process required to prepare these medications from petroleum.

Special-formula arthritis tablets are advertised as providing powerful pain relief. Examine their chemical composition and compare them to regular and extra-strength aspirins. Even stronger pain relievers are available only by prescription. How do these chemical compounds operate in the body?

Conduct a survey in your school or community to determine the average daily use of over-the-counter pain relievers. Does one segment of the population depend on them to a greater extent? Prepare a questionnaire to determine how much people under-stand about the chemical nature of aspirin.

*** Powerful chemical compounds known as endorphins have recently been identified. Produced by the brain, endorphins func-tion as analgesics, or painkillers. Mechanisms controlling their secretion are poorly understood. Some people (for example, yogis who can control involuntary bodily functions to some extent) have been shown to increase endorphin production under certain cir-cumstances. Can you design a research project to investigate the chemistry of endorphins?

*** The pain caused by arthritis is often treated with aspirin. Actually, arthritis is a general term for a family of diseases, includ-ing rheumatoid arthritis, osteoarthritis, and gout. Both the cause and cure are unknown. You may be interested in investigating the chemical basis of arthritic conditions, possibly looking for an effec-tive compound to reduce the high level of uric acid that is responsi-ble for gout.

6 The Bases Are Loaded! Relief Is on the Way!

Was it that fifth hot dog you ate at the baseball game that did it? Maybe it was that extra-large piece of cake after dinner? Or, it may have even been those two super hamburgers you had at the fast-food restaurant for lunch. Whatever it was, you have probably suffered "acid indigestion" at one time or another from eating too much food. To relieve your stomach distress, you may have chewed an over-the-counter tablet, swallowed a chalky liquid, or drunk a bubbly solution. Commercial products sold to relieve upset stomachs are known as *antacids*.

An antacid is a weak base used to neutralize the excess acid occasionally secreted by your stomach. To carry out its digestive role, your stomach contains acidic solutions with a pH value between 1 and 2, by far the most acidic solution in your body. Without the acid, digestive enzymes would not be able to function properly. However, indigestion can result from an overproduction of acid in the stomach, caused either by overeating, stressful conditions, or emotional situations. Antacid products neutralize this excess acid and inactivate the digestive enzymes to calm the stomach and return the acidity to normal levels.

But is it possible for a base to neutralize all the excess acid? One way to find out involves an experimental procedure known as a *titration*. Titration is the process of adding measured amounts of a solution of known concentration to a measured volume of a solution of unknown concentration. You may have performed a titration to determine the concentration of vitamin C in various solutions in Chapter 2. When conducting a titration, an *indicator* is required. Indicators are solutions that change color at specific pH

83

values. In the following experiment, you will use phenolphthalein as an indicator. Vinegar, which is dilute acetic acid, can be neutralized by the addition of a basic solution. You can check various household solutions to determine which is the strongest base by comparing their ability to neutralize the acetic acid in vinegar. In performing a titration, you must arrive at an end point where the indicator changes color.

Have You Reached the End of Titrating?

1. Place 25 ml of vinegar in a clean, dry flask. Add five drops of phenolphthalein and stir. What color is the solution?
2. Fill a buret with a basic household solution such as ammonia. If a buret is unavailable, accurately measure a specific volume of a basic solution and place it in a clean, dry beaker. If you are not sure whether a household solution is basic, check its pH value with universal test paper.
3. While swirling the flask, carefully add the base a drop at a time

Figure 6–1. In performing a titration, the solution in the flask must be swirled after the addition of each drop from the buret until the desired end point is reached.

as pictured in Figure 6–1. Continue adding the base until a noticeable color change is visible. Make sure that the color change remains stable for at least one minute after swirling the flask. Once you have reached the end point, determine the volume of base used to titrate the vinegar. If you did not use a buret, carefully pour the remaining base into a graduated cylinder. As accurately as possible, measure the volume and calculate the volume used by subtracting the remaining volume from the amount originally placed in the beaker.

4. Repeat the titration with another basic household solution. You can use baking soda dissolved in water. Although baking soda is a salt, it forms a basic solution when dissolved in water. Which of the solutions tested is the strongest base? What is the reason for your selection?

5. In place of vinegar, you can try other acidic household solutions, including the juices from citrus fruits, apple juice, and carbonated sodas. You can also experiment with different indicators, such as phenol red, bromothymol blue, and methyl orange. What colors do these indicators turn in acidic and basic solutions?

Extra for Experts

The actual concentration of the acetic acid in vinegar can be calculated from the results of a titration. The concentrations of both acids and bases are usually expressed in *normality* when performing a titration. The advantage is that equal volumes of solutions with the same normality are equivalent. For example, 50 ml of an acid with a normality of 1, expressed chemically as 1N, will be completely neutralized by 50 ml of any 1N base. In fact, 25 ml of a 2N base would effectively neutralize this acidic solution. The equation states that the normality of the acid times the volume of acid is equal to the normality of the base times the volume of the base.

With this in mind, determine the normality of the acetic acid in the vinegar by titrating with a base of known normality. If you prefer, you can type the following program into your computer to calculate the normality. After entering the volume of vinegar placed in the flask, the normality of the base, and the volume of base added to bring about a stable color change, you will be given the normality of the acetic acid in vinegar.

Normality

```
10 PRINT "HOW MANY ML OF VINEGAR WERE USED";:INPUT V
20 PRINT "WHAT IS THE NORMALITY OF THE BASE";:INPUT N
30 PRINT "HOW MANY ML OF BASE WERE ADDED";:INPUT M
40 PRINT
50 A = (N*M)/V
60 PRINT "THE NORMALITY OF THE ACETIC ACID IS ";A;"."
```

Since this is a relatively simple program, you may wish to modify it. By examining some of the computer program listings in previous chapters, can you have this program print your name? What would you add so that the user is asked for the normality of the acetic acid and then is informed whether he or she is correct? How could you modify this program so that you would have only two attempts to answer before being told the correct normality?

There Are More Than Four Bases in Chemistry

All bases share some common chemical characteristics; they have pH values higher than 7 and can react with acids in a neutralization process. In addition, basic solutions can be identified by their slippery feeling. Since bases can cause skin burns, do not use this method to determine if a solution is basic. Instead, rely upon the easiest way to recognize most bases—by their name. The chemical name for a base usually starts with a metal and ends with the word *hydroxide*. Examine the list of ingredients on household solutions you used to titrate the vinegar in the preceding experiment. Can you identify the base present in any of these solutions? Sodium hydroxide, calcium hydroxide, and aluminum hydroxide are chemical names you may discover listed on these labels.

Although bases have similar chemical properties, differences do exist. In fact, some compounds that form basic solutions when dissolved in water cannot be identified as bases by examining their chemical names. For example, sodium bicarbonate and calcium carbonate, whose names obviously do not end with the word *hydroxide*, form basic solutions when dissolved in water. In addition, bases will react differently with other chemical compounds and solutions, depending upon the metal present. Recognizing these chemical differences among bases can be helpful in selecting an antacid brand. All antacids contain a basic compound; however, not all drug manufacturers use the same compound.

In the following experiment, you will have the opportunity to explore the chemical differences between bases. You will investigate such properties as their ability to dissolve in water, their relative strength as measured by pH values, and their reactions with various compounds.

Base Your Answer on the Reaction

1. Working with several bases, place 0.5 gm of each one in a separate test tube. Be sure to label each test tube to keep track of the base it contains. Remember to be careful since bases can cause skin burns. Basic compounds you can use include sodium hydroxide, calcium hydroxide, aluminum hydroxide, sodium bicarbonate, and calcium carbonate. Add 20 ml of distilled water to each test tube. Stopper each test tube and shake vigorously for three minutes. Record the extent to which each base dissolves in the distilled water. Do all the bases dissolve to the same extent?

2. Determine the pH value of each solution by dipping a piece of universal pH paper in the test tube. Which solution is the strongest base? Which one has a pH value closest to 7?

3. Proceeding with one base at a time, equally divide the 20 ml of solution into 4 separate test tubes. To one tube, add 10 ml of 1 percent zinc nitrate solution; to the second, 10 ml of 1 percent ferric sulfate solution; to the third, 10 ml of 1 percent magnesium chloride solution; to the fourth, add 10 ml of 1 percent cobalt chloride solution. Stopper the test tubes and shake to mix the solutions. Record your observations. Look for color changes, formation of a precipitate, production of gas bubbles, or any other evidence that a chemical reaction has occurred (see Figure 6–2). Do all the bases react the same way with any one solution? From your observations, would you say that all bases are chemically identical? Why or why not?

Which Antacid Provides the Best Relief?

With so many different brands of antacids available, a consumer can be easily confused and even misled by advertising claims. In fact, someone suffering from stomach distress may se-

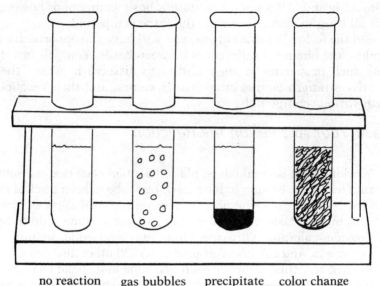

no reaction   gas bubbles   precipitate   color change

Figure 6–2. If no visible changes occur after mixing two solutions, then a chemical reaction may not have occurred. However, the production of gas bubbles, formation of a precipitate, or change in color indicates that a reaction has taken place.

lect an over-the-counter antacid that will do more harm than good. A quick glance at the label on the package reveals that some antacids can cause diarrhea, while others may produce constipation. Some antacids carry warnings for people on low-salt diets or for those suffering from kidney trouble.

Whichever brand is chosen, be aware that antacids should be taken only for temporary relief from acid indigestion. The Food and Drug Administration (FDA) has warned that the recommended dosage for "relief of heartburn, sour stomach, and acid indigestion" be taken for no longer than two weeks since the chemical ingredients in antacids can be harmful if taken for too long. Moreover, if acid indigestion persists, it may be symptomatic of a more serious medical problem. Repeated or chronic episodes of stomach distress may be caused by an ulcer, hiatus hernia, or possibly cancer. Especially when accompanied by severe cramps or vomiting, chronic stomach distress requires prompt medical attention.

But for that occasional bout of acid indigestion, antacids can provide relief. Antacid products have been evaluated for their effectiveness in laboratory analyses. Unfortunately, the results are generally not made known to the public. Consumer-oriented publications occasionally report on their investigations of over-the-counter medications, including antacids. In the following exercise, you will have the opportunity to rate the effectiveness of antacid products. Before testing antacids for their ability to neutralize an acidic solution, first check for the actual base used by various manufacturers.

Although antacid products may contain a combination of ingredients, the base added as a neutralizer is usually one of the following: sodium bicarbonate, calcium carbonate, aluminum hydroxide, or magnesium hydroxide. Examine the list of ingredients on several different antacid products, including both tablet and liquid forms, for the basic compound they contain. Compile a list, indicating both the brand name and the base present. Examine your list with respect to the following information.

Antacids containing sodium bicarbonate are the most potent and can readily neutralize acid. Because of their powerful neutralizing ability, the American Medical Association does not recommend any antacid brand containing sodium bicarbonate. Because it is readily absorbed by the body, sodium bicarbonate can cause *alkalosis,* which is an increase in the blood pH level. Changing the acid–base balance in the blood can lead to serious medical problems. In addition, the sodium present in this type of antacid can be hazardous to people with high blood pressure or kidney impairment. By the way, you can make this type of commercial antacid by simply dissolving half a teaspoon of baking soda in a small glass of orange juice. Baking soda is sodium bicarbonate.

Recognizing the potential health hazard posed by taking this type of antacid, some manufacturers have substituted calcium carbonate in place of sodium bicarbonate in preparing their products. Although it is a powerful neutralizer and also fast-acting, calcium carbonate may cause constipation if used over a long period of time. By adding magnesium salts, which have a laxative effect, manufacturers reduce this potential problem posed by antacids containing calcium carbonate. If taken in large amounts, calcium carbonate can also promote kidney stones.

Antacids containing aluminum compounds are safer than calcium carbonate for people who have problems with either their kidneys or their blood circulation. However, aluminum products can interfere with the body's absorption of important chemicals, including certain antibiotics and the phosphorus needed for healthy bones. Research studies have also suggested that long-term consumption of aluminum-containing antacids can be toxic to the brain.

Magnesium hydroxide, more commonly known as milk of magnesia, is not as powerful a neutralizer as some of the other basic compounds used in antacids. If taken too frequently by people with kidney trouble, magnesium hydroxide may lead to permanent damage. On the positive side, this base is fast-acting and long-lasting.

As you can readily see when comparing your list with the preceding information, the obvious conclusion is that no matter what the active ingredient is, you should never use any antacid brand for a prolonged period of time. The best thing to do is to prevent acid indigestion in the first place. If you are prone to stomach distress, avoid large meals, don't eat hurriedly or under stress, and stay away from foods known to promote acid indigestion. These offenders include coffee, fatty foods, and chocolate.

How Effective Is an Antacid?

With millions of dollars spent each year on antacid products, are most people getting their money's worth, especially when they pay a little extra for a "special formula" or "fast-acting" brand? To determine which antacid is the most effective in neutralizing excess acid, use the following procedure developed by the FDA for analyzing over-the-counter products. Although the FDA-recommended procedure calls for the use of a pH meter, several indicators are suitable for this analysis.

1. Dissolve one antacid tablet in 70 ml of distilled water in a clean, dry flask. If the tablet is a chewable type, first crush it before adding the water. Then add 30 ml of 1N hydrochloric acid and stir. Allow the contents to stand for 15 minutes with occasional stirring.
2. If a pH meter is unavailable, add 10 to 15 drops of either bromophenol blue or congo red indicator. Record the color. If the antacid tablet is artificially colored, you will have to add enough indicator to produce a noticeable color.
3. Fill a buret with 0.5N sodium hydroxide solution. If a buret is not available, pour a measured volume of the sodium hydroxide solution into a clean, dry beaker. While stirring, add the base drop by drop until a stable color change persists for at least one minute. If you are unsure whether a definite color change has occurred, add a few drops of the indicator first to hydrochloric acid and then to sodium hydroxide. The end point of the titra-

tion is reached when the indicator produces the same color as it does when mixed with the sodium hydroxide. If a pH meter is used, add the sodium hydroxide until the pH stabilizes at 3.5 (see Figure 6–3). Record the volume of base added to the ant-acid–acid solution. If a beaker was used, pour the remaining sodium hydroxide solution into a graduated cylinder and deter-mine the volume of base added to the flask.

4. The FDA requires antacid manufacturers to rate the effective-ness of their product in terms of the amount of acid neutralized by the base. The amount of acid neutralized is expressed in *milliequivalents*. Calculate the milliequivalents of acid neutral-ized by the antacid tablet by placing your results in the follow-ing equation:

milliequivalents of acid neutralized = (30 ml, the volume of acid used) × (1, the concentration of the acid) minus (the ml of base added) × (0.5, the concentration of the base)

For example, if 50 ml of base were added to bring about the color change, then the milliequivalents of acid neutralized is calculated as follows:

$$(30) \times (1) - (50) \times (0.5),$$

that is, 30 minus 25, which equals 5 milliequivalents of acid

5. To test the effectiveness of liquid antacids, pour an amount

Figure 6–3. A small, inexpensive meter can be used to measure the pH of solutions more rapidly than testing with universal indicator paper.

equivalent to the recommended dosage into a graduated cylinder. Add distilled water to bring the volume to 70 ml and then pour this liquid into a clean, dry flask. Add 30 ml of 1N hydrochloric acid and stir. Allow the contents to stand for 15 minutes with occasional stirring. Continue the analysis, beginning with step 2.

6. Prepare a bar graph similar to that in Figure 6–4 summarizing your findings. Along the horizontal axis, indicate the brand name; along the vertical axis, record the milliequivalents of acid neutralized either per tablet or per liquid dosage. According to your graph, which antacid brand is most effective? Which one is the least effective? If you compared a "regular" and "special-formula" antacid produced by the same manufacturer, is there a significant difference between the two? If not, can you explain why the manufacturer claims that one is a "special-formula" brand? What is meant by a manufacturer's claim that an antacid "consumes 47 times its weight in excess stomach acid"?

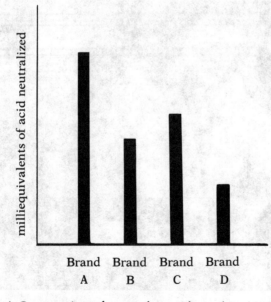

Figure 6–4. Constructing a bar graph provides a clear, pictorial representation of experimental results.

Topics for Further Investigation

Several antacid manufacturers add a special ingredient called simethicone to "reduce gastrointestinal gas." Because of this additional compound, these drug companies charge more for their products. However, many chemists question the effectiveness of simethicone. Can you design an experiment to test whether simethicone reduces gases produced in the stomach and small intestine?

Check the labels on various brands of antacids for chemical ingredients other than basic compounds. What purpose does each of these additives serve? Can any of these ingredients be potentially hazardous to someone's health?

The function of bases in antacids is to neutralize excess stomach acid. Investigate other chemical and industrial uses for basic compounds.

In addition to its use in titration analyses, what other important and practical applications does the pH scale have? Examine its use in testing water samples and its role in chemical reactions conducted by industrial manufacturers.

If you have discovered that a manufacturer's advertising claim or package labeling was erroneous or misleading, contact the company asking for an explanation of their position. Be sure to include the results of your analyses to support your position.

If a base completely neutralizes an acid, a salt and water are produced. Compile a list of salts prepared by neutralization reactions that are useful to consumers. Begin by investigating salts used in medicines, fertilizers, and paints.

Develop arguments both for and against passing a law requiring antacid manufacturers to indicate the effectiveness of their product on the package label.

*** Proteins circulating in your blood serve as buffers to prevent any change in the blood pH. Their importance can be appreciated when it is recognized that even a slight shift in the blood's pH (normally 7.4) can result in serious illness or death. Investigate the chemical action of these protein buffers.

*** Neutralization reactions result in the production of a salt and water. Chemists are experimenting with different salts for a variety of purposes, including their use in air conditioners. Salts that freeze at relatively low temperatures are especially valuable. As the temperature increases during the day, these salts would melt, absorbing the heat and thereby reducing the amount of electricity required by the air conditioner. Can you develop a salt with such special properties?

7   It's Time to Get Ready for School

It's seven o'clock in the morning as your alarm clock buzzes. You hop out of bed, jump in the shower, and pour some sodium lauryl sulfate on your hair. After you dry off, you may rub your hands with propylene glycol monostearate, put a little sodium borate on your face, place a dab of calcium carbonate in your mouth, or splash your face with isopropyl alcohol. Now you're attractive and ready for school!

Actually, all you may have done in preparing for school is to shampoo your hair and then soften your hands with a lotion, clean your face with cold cream, brush your teeth, or put on some aftershave lotion. These cosmetic products, used by millions of people every day, are made from chemical compounds. Yet, long before chemists developed the specialized procedures that would become the basis of the cosmetic industry, people used cosmetics for a variety of purposes. In ancient times, physicians prepared and dispensed cosmetics primarily to women seeking to enhance their appearance. As the popularity of cosmetics grew, these products began to be used by an increasing number of men; today almost as many cosmetics are sold to men as to women. From its simple beginnings, when physicians mixed a few chemicals, the cosmetic industry has developed into a major business, employing highly trained chemists who constantly search for new formulas.

Billions of dollars are spent annually on cosmetics, which are among the most heavily advertised products in the world. Lured by advertising claims, people buy them to smell nice or not smell; to straighten hair or make it curl; to dry skin or add moisture to it.

In fact, a cosmetic is defined by law as any product intended to clean, beautify, or change one's appearance. Many people spend a considerable amount of money on a cosmetic product, believing that it will perform wonders. However, unknown to most consumers, the cost of the chemical ingredients used in the manufacture of a cosmetic is a fraction of the retail price. Most of what you pay covers advertising costs and goes to the manufacturer's and retailer's profits.

Many cosmetic products are either solutions, suspensions, or emulsions of a few standard ingredients. For example, all soaps and shampoos contain a special chemical compound to break up and disperse the oils and grease that trap dirt on the skin or in the hair. Washing the skin or hair without soap or shampoo is not effective, because the oils and trapped dirt do not mix with water and therefore cannot be easily removed by just rinsing with water. The chemical ingredient in shampoos acts as an emulsifier by surrounding the oil droplets and causing them to disperse or spread throughout the water (see Figure 7–1). Once the oils and grease are dispersed by the emulsifier, the dirt can be more easily washed away with water.

Although all shampoos contain a chemical to disperse oils and grease, each manufacturer may have its own special combination of ingredients. Often kept secret from both competitors and consumers, these special formulas for making shampoos may be used to make the product have a "natural pH," "extra body," or "power to heal split ends." But is a special-formula shampoo more effec-

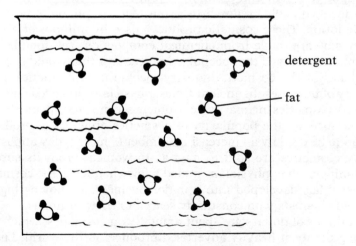

Figure 7–1. Fat molecules, surrounded by detergents, can be emulsified, or evenly distributed, throughout a water solution.

tive in cleaning hair? In the following investigation, you will have the opportunity to explore several basic properties of shampoos. You may purchase your next bottle or tube of shampoo based on the results of this study!

How Good Is Your Shampoo?

1. Prepare a 1 percent solution of various shampoo brands by diluting 2 ml of the shampoo with 198 ml of distilled water. Using universal pH paper, determine the pH value for each shampoo. Which brand has a pH value closest to neutral? What do you think is meant by the phrase "natural pH"? Can this phrase be applied to any of the shampoos? Good shampoos have a pH close to neutral, while those whose pH values are acidic can damage hair.

2. Pour 25 ml of each 1 percent shampoo solution from step 1 into a 100-ml graduated cylinder. To avoid producing any foam, pour the shampoo down the side of the graduated cylinder as shown in Figure 7–2. Stopper the cylinder or cover tightly with your palm and shake the cylinder vigorously ten times. Record the volume of foam produced. A good shampoo will produce at least twice its volume in suds when shaken ten times. Consequently, 25 ml of shampoo should give 50 ml of foam, or a 2:1 ratio of foam to liquid. Which brand tested gives the highest ratio of foam to liquid? Proceed as quickly as possible to the next step.

3. Remove the stopper from the graduated cylinder. Record the volume of foam remaining at the end of each minute. Do this for five minutes. Graph your results by plotting the volume of foam on the Y axis and time in minutes on the X axis. Based on your graph, which brand of shampoo retains its foam for the longest time?

4. Pour another 25 ml of each 1 percent shampoo solution from step 1 into a 100-ml graduated cylinder. Add one drop of india ink, stopper tightly, and shake ten times. Determine if the india ink has dispersed in the foam, liquid, or both. Which brand tested has the highest amount of ink in the foam? in the liquid? Good shampoos will disperse the ink in the liquid. Solid materials should remain suspended in the liquid fraction so that they can be washed away with water. Dirt and grease trapped in the foam are more difficult to rinse away with water.

5. The "body" of a shampoo is known as its *viscosity*. The higher the viscosity, the "thicker" or more concentrated the shampoo. To measure the viscosity of shampoo, fill a 100-ml graduated

Figure 7–2. Pouring the shampoo, or any solution likely to foam, down the side of the cylinder will eliminate the formation of suds that can interfere with volume measurements.

cylinder with shampoo. Determine the time it takes for a small marble or steel ball bearing to fall to the bottom of the graduated cylinder. Which shampoo has the greatest viscosity, or the "most body"?

6. Calculate the cost per 5 ml of shampoo. Which brand is the most expensive? the cheapest? Based on your results, is there one brand of shampoo you would now make it a point to buy?

Get Your Hands into This Emulsion!

All hand lotions are made by mixing an oil with water. As you realize from your study of shampoos, oil and water do not mix. Consequently, the manufacturer must add a third ingredient to act as an emulsifier to disperse the oil in water. In addition, the manufacturer usually includes a chemical to prevent the growth of bacteria and some type of perfume to mask any unpleasant odors produced by the chemical ingredients. A variety of oils can be used in the preparation of hand lotions, including fatty acids, lanolin, mineral oil, or the oils from olives and peanuts. Water makes up most of the hand lotion, while glycerol is commonly used as the emulsifying agent.

When rubbed onto the skin, hand lotions containing glycerol evaporate, producing a cooling effect. The lotion also moisturizes, because a small amount of glycerol remains as a thin film on the surface of the skin, preventing the evaporation of water. In the following exercise, you can prepare an inexpensive hand lotion and then compare it to commercial brands. Since you have the opportunity, you can experiment with different perfumes to produce the best-smelling hand lotion.

1. Make sure all glassware used in the preparation of your hand lotion is thoroughly cleaned. Even a small amount of an impurity may cause skin problems.
2. Mix 2 gm of boric acid with 20 ml of glycerol in a beaker. Gently heat these ingredients to 70°C while constantly stirring with a glass rod until the boric acid dissolves. The boric acid serves as the antibacterial agent.
3. Allow the beaker to cool and add 12 gm of lanolin and 28 gm of petrolatum, also known as petroleum jelly. The amount of perfume added depends upon your personal preference. Stir gently while heating until the mixture liquefies. Allow to cool overnight. You can store your hand lotion in a clean bottle.
4. Experiment by varying the concentrations of the ingredients to see how the viscosity or "body" of the lotion is affected. How does the hand lotion you made compare with commercial products? In the following analysis, you will put your product up against the leading brands.

Try Your Hand at Analyzing Lotions

1. Rub a few drops of your lotion into your hands. Determine its effectiveness in softening the skin, the feeling of coolness it produces, and its fragrance. You can score each category with 0, +, ++, or +++. If no softening effect can be detected, then the lotion would receive a 0 rating, whereas one producing a very noticeable difference would get a +++. Compile a table indicating the scores for each of the three categories for all the lotions tested.

2. To perform a less subjective evaluation of the softening efficiency of each hand lotion, soak a small piece of dialysis tubing
in water for a few minutes. Dialysis tubing is a synthetic mem-
brane that is similar to the membranes surrounding your skin
cells. Remove the tubing and shake off any excess water. Apply
a very thin coating of hand lotion to cover the tubing. Place the
tubing under a heat lamp or incandescent light and note the
time required for the tubing to dry. Which lotion was most
effective in preventing evaporation of water?

3. Mix a small amount of each lotion with 10 ml of distilled water
in a test tube. After thoroughly mixing the contents, check the
pH with universal test paper. Does one brand have a pH value
closest to neutrality? Could any of them cause a skin problem
because of its pH value?

4. Rate the "thickness" of each lotion (see Figure 7–3) by checking
its viscosity as you did with the hair shampoos. Which lotion
has the most "body"?

5. You can verify your results from the preceding step by determining the solid content of each hand lotion. The more chemical compounds present, the more concentrated the lotion. Place
5 gm of each lotion into an evaporating dish which has been
cleaned and weighed. Gently heat the lotion until only solid
material remains. Weigh the dish and solid. Subtract the
weight of the dish from that of the dish and solid material to
determine the weight of the solid. Determine the percent of
solid material in each lotion sample by placing your results in
the following equation:

$$\frac{\text{weight of solid material}}{\text{5 gm (weight of lotion)}} \times 100 = \text{percentage of solid material}$$

6. Calculate the cost per 5 ml of lotion. Determine the cost of the
chemicals used in preparing your lotion. Assuming the cost of
the ingredients to be approximately the same for the manufacturer, what percent of the retail price covers the cost of the
chemicals used to prepare one of the commercial lotions you
tested?

A Cream to Clean

Cold or cleansing creams are among the first cosmetic products used by women to remove oily deposits and the grease left by
makeup. Like hand lotions, cold creams are emulsions prepared by

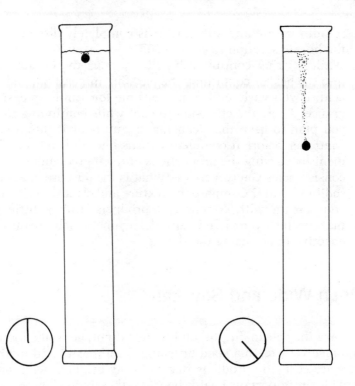

Figure 7–3. The time taken for a small ball bearing to fall through the solution will depend upon the shampoo's viscosity.

adding an emulsifying agent to an oil–water mixture. Varying the proportion of these ingredients can change the consistency of the emulsion, producing anything from a liquid to a thick cream. When applied to the skin, the components of the emulsion separate. The water evaporates, producing a cooling effect on the skin, while the oil dissolves any grease or dirt, which can then be removed by wiping with a tissue or towel. In the following exercise, you can prepare a cleansing cream similar to commercial products.

1. Place 20 gm of paraffin which has been broken into small pieces into a large beaker. Heat until all of the paraffin has melted. Be careful not to heat too strongly or the paraffin will burn.

2. Add 125 ml of mineral oil and 1.5 gm of stearic acid to the melted paraffin. Maintain the temperature of this mixture between 60°C and 70°C.

3. Heat 70 ml of water to 70°C in a second beaker. Add 1.5 gm of

sodium borate and stir until it is completely dissolved. Allow the borate solution to cool to 60°C.

4. While stirring continuously, slowly pour the borate solution into the beaker containing the paraffin, mineral oil, and stearic acid. If you wish, add some perfume for a mild, pleasing fragrance. Allow the emulsion to cool while continuing to stir. If you plan to keep the cleansing cream, pour it into a suitable container before it completely turns into an emulsion. Experiment by varying the proportions of the ingredients to see what consistencies you can create. What is the purpose of adding the sodium borate? Compare the texture and cleansing efficiency of your cream with commercial products. Rub a little greasy makeup into your hands and determine which cream is most effective in cleansing the skin.

Open Wide and Say Aah

Are they clean, bright, and white? If not, get some toothpaste and give your teeth a good brushing. Toothpastes and tooth powders are cosmetic products that not only brighten teeth but also provide an important health benefit—they reduce the possibility of tooth decay. With the increased amount of sugars and soft foods in our diets, the incidence of tooth decay has risen drastically in many populations. In fact, tooth decay is the most common disease in the world, affecting almost every individual.

Tooth decay is caused by a sticky, almost invisible substance called *plaque* that forms especially around the gum line. Plaque develops when bacteria in the mouth act on sugars, changing them into acids that destroy the enamel surface of teeth. Although enamel is the hardest substance in your body, the acids secreted by the bacteria can erode this protective covering, leading to a cavity. Brushing with toothpaste aids in the removal of plaque found on the surfaces of teeth. Flossing removes plaque that collects between teeth and near the gums. In the following investigation, you will examine several properties of toothpastes.

Good Toothpaste Is Abrasive

1. To remove plaque, a toothpaste must have some abrasive action, which is the ability to remove something with the fric-

tion produced by rubbing. Determine the abrasive power of the toothpaste by placing a small amount of various brands on different clean, unmarked glass microscope slides. Place another clean slide on top and gently rub the two slides together 50 times as pictured in Figure 7–4. Remove the top slide, wash it with water, dry it with lens cleaning tissue, and then examine it very carefully under a microscope for scratches. Record the abrasive action of each toothpaste brand as either light, moderate, or heavy. Based on your results, is one brand more effective in removing plaque?

2. A chemical compound often used as an abrasive is calcium carbonate. To test for its presence, place 3 gm of the toothpaste in a test tube and add 6 ml of 1N hydrochloric acid. If calcium carbonate is present, a gas will be released, as evidenced by the production of small bubbles. Seal the test tube with a one-hole rubber stopper fitted with a small piece of glass tubing. Attach a small length of a rubber hose and place the other end in a second test tube containing limewater (see Figure 7–5). Allow the gas to pass through the limewater. If carbonate is present,

Figure 7–4. Rubbing toothpaste between two glass slides will produce scratches on the slides that can then be examined under a microscope. The more scratches, the more abrasive the toothpaste.

the solution of limewater will turn cloudy.
3. The cleansing action of the toothpaste can be examined by placing 1 gm into a 100-ml graduated cylinder and adding 50 ml of
   water. Place your hand over the top of the cylinder and shake
   vigorously for one minute. Record the volume of foam produced. A good toothpaste will foam easily and produce a sufficient quantity for cleansing action.
4. Place a small piece of universal pH paper in your mouth. Record the pH value. Place a small dab of toothpaste in 10 ml of
   distilled water and shake to disperse the sample. Dip a small
   piece of universal pH paper into the toothpaste solution. Which
   brand has a pH value that is closest to that of the saliva in your
   mouth?

Let's Mix Up Some Toothpaste

Toothpaste is a *mixture*. Whenever two or more substances are
mixed together but each still retains its original properties, then a
mixture has been prepared. For example, adding salt to water
produces a mixture of salt water. How would you prove that both
the salt and water retain their chemical properties even after they
have been mixed? Is there some way to separate the salt from the
water so that each of these two chemicals can be shown to have
retained its original properties? Mix the right ingredients, and you
can make toothpaste. In the following exercise, you can prepare
some homemade toothpaste.

1. Place 8 gm of powdered sugar (for that sweet taste) in a mortar.
   Add two drops of oil of peppermint (for that mint flavor) and 3
   gm of castile soap (for that cleansing action). Mix these ingredients with a pestle.
2. Add 22 gm of calcium carbonate (for that abrasive action) and
   mix thoroughly. Then add corn syrup or glycerol and mix until
   a paste with a uniform consistency is produced. Can you figure
   out how manufacturers get the paste into that little hole in the
   tube?

Try This Solution

After-shave lotions are prepared by mixing various chemicals
to form a solution. A good after-shave lotion should smell pleasant
and dry quickly after it has been splashed on your face. A few
minutes is all that is required to prepare the lotion. Mix 30 ml of

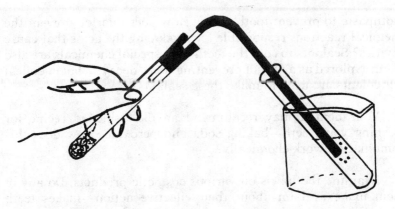

Figure 7–5. The gas produced by reacting toothpaste with 1N hydrochloric acid is passed through a solution of lime water to determine if calcium carbonate is present in the toothpaste.

isopropyl alcohol, 30 ml of distilled water, 5 ml of glycerol, 1 gm of menthol, and a few drops of perfume in a clean beaker. Experiment by varying the proportions to make a product better suited for oily skin. Splash it on your face, and you're ready for school!

Topics for Further Investigation

Trace the historical development of the cosmetic industry from biblical times to the present. What impact has this industry had on people? Find out what education, experience, and expertise are required to become a research chemist for a cosmetic manufacturer.

In this chapter you have investigated the chemical basis of several cosmetic products. There are, however, numerous cosmetics that have not been examined, including shaving creams, mouthwashes, hair conditioners, lipsticks, makeups, hair dyes, nail polishes, colognes, perfumes, foot powders, and deodorants. Contact a manufacturer to obtain information on the chemical nature of these products.

With the problem of tooth decay affecting nearly everyone in the world, scientists are searching for preventative measures, including more effective toothpastes. Fluoride is often added to toothpaste to prevent tooth decay. How does fluoride prevent the chemical reactions responsible for producing the acids that cause cavities? Sealants to coat the teeth with special chemicals are also being explored as a way of preventing tooth decay. What chemical ingredients are used to make these sealants?

Your dentist may recommend an old-fashioned recipe for cleaning your teeth—baking soda and peroxide. How does this combination work chemically?

Examine the labels on various cosmetic products. Do any of them make a claim about their effective action—makes teeth whiter, removes aging skin spots, etc.? Design an experiment to check the truth of the advertising claim. You may want to inform the manufacturer of your results.

Are cosmetic products used to the same extent all over the world? An interesting social studies report could center on the attitudes, practices, and customs of people in different countries toward cosmetic products.

Conduct a survey in your school or community to determine the average number of different cosmetic products used daily by each person. Is there a significant connection between cosmetic use and one sex or age group?

Have any cosmetic products been banned by the Food and Drug Administration? If so, what was the reason for the FDA's action?

*** Select a cosmetic product and carefully research its chemistry. Modify the chemical procedure or develop an entirely new one to make a product that is more effective, yet reasonably inexpensive.

*** Investigate the chemical action of emulsifiers and explore modifications to make them more effective agents for manufacturing cosmetics.

CHEMISTRY IN THE LAUNDRY ROOM

8 Soaps and Detergents

Washing with soap and water is one way to get rid of that "ring around the collar" on your shirt or blouse. The cleansing action of soap involves the dissolving of the oils and grease that bind dirt to the surfaces of clothes. As you may recall from your study of emulsions, water and oil do not mix. Adding an emulsifying agent disperses the oil in water. Because of its structure, the chemical compound in soap acts as an emulsifier. One end of the compound dissolves in water, while the other end dissolves in oil or grease. By dissolving in water at one end and in oil or grease at the other end, the compound in soap can emulsify or separate them. The dirt particles, no longer trapped by the grease, can then be washed away with water.

Soaps not only disperse grease in water but also help to separate the particles or molecules of water. A *molecule* is the smallest particle of any substance that retains all its chemical characteristics. As a result of their strong attraction for one another, water molecules stick close together, especially along a surface. This strong attraction between water molecules on a surface is called *surface tension*. The high surface tension beteeen water molecules causes them to form a thin layer, or "skin," capable of supporting a lightweight object such as an insect or pin. To verify this phenomenon, observe a small insect walk on water. To demonstrate the surface tension of water, use a tweezer to float a pin or some other lightweight metal object on a bowl of water. Without hitting the pin, carefully add liquid soap drop by drop. Explain your observations.

107

Soap from Ashes and Fat

Although they were unaware of the chemical structure and action of the compound in soap, people have followed recipes for making soap for centuries. The early Romans mixed the ashes from burned wood with the fat from sacrificed animals to form a crude soap. As its value for cleansing purposes became better known, soap began to be more widely used. In Europe during the Middle Ages, soap was prepared by combining the ashes from trees with the fat from goats. The demand for soap, especially by the rich and by the nobility, turned soap making into an art. With the development of chemistry as a science in the eighteenth century, chemists discovered simpler, cleaner, and cheaper ways to manufacture soap.

Known as *saponification,* soap making involves the reaction of a fat with a base, usually sodium hydroxide or potassium hydroxide. Although all soap manufacturers depend upon a fat and a base as their starting point, they may vary the proportion of these ingredients for certain purposes. If a harder soap is desired, more sodium hydroxide is used; to make a softer soap, the concentration of potassium hydroxide is increased. Recently, manufacturers have modified the saponification procedure to produce "liquid soap" that can be pumped from a dispenser. In addition, the manufacturer may include chemical additives to improve the quality, purity, or smell of the soap. See what type of soap is made in the following procedure, which is similar to the recipe your great-grandmother probably used to make soap for her family.

Making Soap the Old-Fashioned Way

1. Place 10 gm of lard or animal fat in a 250-ml beaker. Add 15 ml of 6M sodium hydroxide and 50 ml of ethyl alcohol. Be very careful when adding these two chemicals since the sodium hydroxide can cause skin burns and the alcohol is flammable.
2. While stirring with a glass rod, gently heat the beaker over a Bunsen burner or hotplate for at least 30 minutes (see Figure 8–1) to make sure that the base has completely reacted with the lard.
3. Add 20 ml of distilled water and stir thoroughly.
4. While this mixture is cooling, add 12 gm of sodium chloride to 50 ml of distilled water. Pour the cooled mixture of fat and base into the sodium chloride solution. After allowing the contents to cool completely, remove the solid cake. This is the soap.

Figure 8–1. In the saponification reaction, the ingredients must be heated for at least 30 minutes to make sure the sodium hydroxide has completely reacted with the lard.

5. Before using this soap, test it for the presence of any base that has not reacted with the fat. If the soap is too basic or alkaline, it may irritate the skin. To check the alkalinity, dissolve a small piece of the soap in water and determine the pH value with universal test paper. If the pH value is greater than 8, place the soap in a clean beaker and gently heat it until it liquefies. Continue heating for 15 minutes so that any base present will react with the fat. After it cools, remove the soap and again check its pH value.

Is Your Soap "99 and 44/100 Percent Pure"?

Now that you have performed a saponification reaction to make soap, test it to see how good a product you've prepared! In the following investigation, you will examine several chemical

Magnesium sulfate heptahydrate
$MgSO_4 \cdot 7H_2O$

CHEMISTRY AROUND YOU

properties of your soap, including its capacity to foam, its ability to reduce surface tension, and its emulsifying power. But before analyzing your soap, first prepare some hard water by dissolving 5 gm of Epsom salts in 1 liter of water. Hard water is characterized as containing a high level of salts and minerals. You will investigate the action of your soap in tap, distilled, and hard water.

1. Place a small piece of your soap in each of three test tubes. Half-fill one test tube with tap water, the second with distilled water, and the third with hard water. Place your thumb over the first test tube and shake vigorously for 30 seconds. Record the height of the suds layer. Examine the soap sample to determine how much has dissolved. Also note if the solution is clear or cloudy (see Figure 8–2). Repeat this procedure with the other two test tubes. In which water sample did the soap form the most suds? Did the soap dissolve to the same extent in all three water samples? Were any of the soap solutions clear or transparent?

2. Dip a small cotton ball in mineral oil and wipe it on a clean microscope slide so that a thin film covers the slide. Using a medicine dropper, place two drops of water on the slide. Record your observations. Cover a second slide with mineral oil. Place two drops of your soap solution prepared with tap water on this slide. Explain your observations.

3. To test the emulsifying action of your soap, half-fill a test tube with distilled water. Add enough corn or olive oil to form a thin

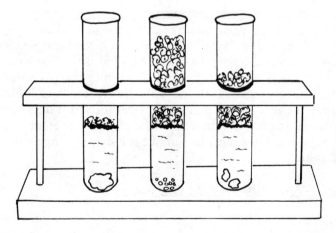

Figure 8–2. Three equal-sized soap samples have been shaken in different types of water. Which test tube contains tap water? hard water? distilled water?

layer on the top of the water. Shake vigorously for 15 seconds and then allow the test tube to remain undisturbed for two minutes. Record your observations. Repeat this procedure, substituting your soap solution made with distilled water. Account for any difference between the result produced by shaking distilled water with the oil and that produced by mixing the soap solution with the oil.

Soap and Drain Cleaner–Something in Common

As you have discovered, the saponification reaction is the chemical basis for making soap. But saponification can be useful in solving a household problem occasionally encountered in kitchens and laundry rooms—clogged drains. If you check the label on either liquid or solid drain cleaners, you'll notice that the main chemical ingredient is a base. When the drain cleaner is poured into a clogged pipe according to the directions on the package, the base reacts with the grease clogging the drain to form a soapy mixture. Pouring water down the drain can then unclog the pipe.

In fact, you can prepare soap by substituting 5 gm of solid drain cleaner for the 15 ml of 6M sodium hydroxide used in the saponification reaction you performed. Exercise caution when using drain cleaners since they can cause serious skin burns because of their strong alkalinity. By the way, the strong base in these cleaners may not be effective if there is a major obstruction in the pipe. In this case, the base may even erode the plumbing if it does not react to form soap but simply remains trapped in the drain.

Detergent—A More Effective Cleaner

As you may have noticed in testing your soap with hard water, a noticeable deposit or residue can form. Chemicals in the soap react with the magnesium and calcium salts found in hard water to produce a *precipitate,* a solid substance not readily soluble in water. A familiar example of a precipitate formed when soap reacts with hard water is the ring left in a tub after bathing. Since the deposit is difficult to remove, clothes may not get their brightest or cleanest after laundering in hard water.

Recognizing that this residue might pose a problem in households where the water is naturally hard, chemists explored for ways to prevent the formation of the precipitate without reducing the cleaning power of soap. They discovered that adding various ingredients to soap would inhibit the formation of the residue when washing in hard water. Their work led to the development of a new type of cleaning compound known as detergents. Since they contain chemicals that dissolve in hard water, detergents do not form an insoluble precipitate. Instead, detergents produce just as many suds in hard water as in soft water. Because of this chemical property, detergents rapidly replaced soaps, particularly in communities where the water is hard.

Although they both act chemically as emulsifiers, detergents differ from soaps in that an alcohol is used in place of fat during the manufacturing process. An alcohol commonly added during the saponification process is lauryl alcohol, which is made from petroleum. By the way, if you're surprised to discover that a cleansing agent is partly derived from a sticky, black substance like oil, you may want to investigate which other home products are manufactured from petroleum derivatives.

Although the use of detergents eliminated "bathtub rings," it introduced more serious problems, affecting not only individual households but also entire communities. The chemical compound first used as the emulsifier in detergents was not *biodegradable,* that is, capable of being broken down by the action of bacteria and other small organisms. When released from sewage plants, these detergents contaminated the water because small organisms living in the rivers, lakes, or streams could not break down or degrade the emulsifier. For this reason, these detergents accumulated in the water, sometimes forming suds and often posing a health hazard to both animal and plant life. To solve this pollution problem, chemists modified the emulsifier in detergents so that it is now biodegradable.

However, the emulsifier was not the only problem encountered when using detergents. A phosphate compound, normally added to detergents because of certain chemical properties, was found to be another threat to aquatic life. Since the phosphates combine with the magnesium and calcium in hard water, they are added as water softeners. By combining with these salts, the phosphates prevent them from forming insoluble precipitates. The phosphates also serve as *buffers* by keeping the pH of the laundering solution from fluctuating over a wide range.

Although phosphates serve to both soften and buffer laundering solutions, they pose a serious pollution problem. Phosphates are not biodegradable and consequently collect in rivers, lakes, and streams, where they serve as a food source for algae, which are microscopic plants. With a rich food supply, the algae proliferate and become the predominant life form in these bodies of water. Using all the available oxygen, the algae make conditions impossible for fish and other organisms to live. In response to consumer activist groups, some detergent manufacturers have completely eliminated phosphates, while others have reduced the levels present in their products. Currently, chemists in the detergent industry are searching for a compound that will soften and buffer, but not pollute.

With an understanding of the chemistry of detergents, how about trying to make your own brand? In the following exercise, you can prepare a detergent and then test some of its properties. In fact, you will have the opportunity to put it up against some of the leading commercial brands to determine which one is most effective as an emulsifier!

Don't Delay! Discover Detergents!

1. Place 15 gm of lauryl alcohol (dodecanol) in a clean 250-ml beaker. Slowly add 10 ml of 6M sulfuric acid to the alcohol. Be very careful when adding the acid since it can cause serious skin burns.
2. Pour 30 ml of 6M sodium hydroxide solution into a second 250-ml beaker. Since the sodium hydroxide is a strong base, be careful when making this solution. Add three drops of phenolphthalein indicator to the sodium hydroxide solution. Is there any color change in the indicator?
3. Carefully pour the alcohol–acid solution slowly and with constant stirring into the sodium hydroxide solution. Describe the reaction.
4. Filter the solution and allow the precipitate to dry on the filter paper overnight. This is the detergent; save it so that you can test its effectiveness as an emulsifier.

Detergent Power

The efficiency of a detergent is determined by its ability to emulsify the grease that traps dirt on clothes. In the following exercise, you will investigate the amount of detergent required to emulsify a fat sample in water. Obviously, the detergent using the least volume to disperse the grease in water is the most effective cleaning agent.

1. To observe the emulsion process more easily, first prepare a colored fat sample by melting some lard and then adding enough Sudan IV dye to obtain a dark pink or light red color. Solidify the fat by refrigerating it overnight. The dye will enable you to observe what happens to the fat as it is mixed with water and detergent.
2. Heat 500 ml of tap water to 55°C in a flask or a beaker. Maintain this temperature as closely as possible throughout the experiment. Add 2 gm of the colored fat to the water. Swirl the flask or beaker and record your observations.
3. Slowly add the detergent to the fat–water sample while continuously swirling the flask or beaker. Continue adding the detergent until the fat sample is completely emulsified, as evidenced by the dispersion of the colored fat in the water—the fat should no longer remain as a distinct globule on the surface of the water (see Figure 8–3). Record the volume of detergent added to emulsify the fat sample.
4. After you have completed the analysis of various brands, calculate their relative efficiency by determining the amount of fat emulsified per milliliter of detergent. Construct a bar graph, indicating the brand name and the amount of fat emulsified per milliliter of detergent. Which brand is most effective? How did

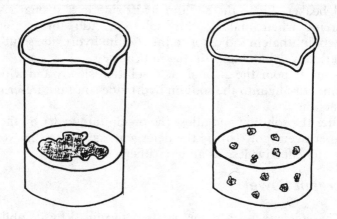

Figure 8–3. Without detergents, fat remains as a distinct glob on the surface of the water. As detergent is added, the fat is emulsified into tiny globules, eventually dispersing throughout the water.

the detergent you prepared compare with commercial products?

5. Add 1 ml of detergent to 9 ml of distilled water to make a 10 percent solution. Check the pH value with universal indicator paper. Which brand has a pH value that would be best for sensitive skin and delicate fabrics?

6. The amount of emulsifier in a detergent can be approximated by reacting it with the dye methylene blue. Add one drop of 1 percent methylene blue solution to 10 ml of tap water in a test tube. Stopper the tube and shake for five seconds. Add 5 ml of hexane or another nonhazardous organic solvent to the test tube. Since the organic solvent and water do not mix, two distinct layers will form. Is the solvent less or more dense than water? In which layer does the methylene blue dissolve?

7. Add ten drops of liquid detergent, stopper the test tube and shake for 15 seconds. After allowing the organic solvent and water layers to separate, observe if any of the dye is dissolved in the solvent. The amount of dye dissolved in the solvent depends on the amount of emulsifier in the detergent—the greater the concentration of emulsifier, the more dye (darker color) in the solvent layer. In which brand is the emulsifying agent most concentrated? Do the results agree with the data obtained in step 4—is the most effective emulsifying detergent the most concentrated?

8. Repeat this analysis with liquid detergents sold for washing dishes to determine their effectiveness in emulsifying the grease that collects on plates and glasses. How would you modify this procedure to test powdered detergents?

 Detergents

The following program will simplify the calculations needed to determine which brand of detergent is the most effective emulsifier. After entering the number of brands analyzed, you will be asked to input the volume of detergent added in each case. After displaying the grams of fat emulsified per gram of detergent, the computer will select the most effective brand. You may want to modify the program if you've analyzed several brands so that the computer screen is cleared after the result for each brand is given.

```
10 PRINT "WHAT IS YOUR FIRST NAME";:INPUT NA$
20 PRINT "HOW MANY BRANDS WERE ANALYZED";:INPUT N
30 PRINT "TYPE IN NAME OF THE FIRST BRAND";:INPUT B$
40 PRINT "HOW MANY ML WERE ADDED";:INPUT B
50 C = B*.1:D = 1/C:C$ = B$
60 PRINT D;" GRAMS OF FAT WERE EMULSIFIED"
70 PRINT "PER GRAM OF DETERGENT, ";NA$;"."
80 FOR X = 1 TO N − 1
90 PRINT "WHAT WAS ANOTHER BRAND TESTED";:INPUT A$(X)
100 PRINT "HOW MANY ML WERE ADDED";:INPUT A(X)
110 A(X) = A(X)*.1:A(X) = 1/A(X)
120 IF A(X) > D THEN C$ = A$(X):D = A(X)
130 PRINT A(X);" GRAMS OF FAT WERE EMULSIFIED"
140 PRINT "PER GRAM OF DETERGENT, ";NA$;"."
150 NEXT X
160 PRINT "THE MOST EFFECTIVE EMULSIFYING"
170 PRINT "DETERGENT IS ";C$;"."
180 END
```

You Need Chlorine to Really Clean

Washing dirty clothes with an emulsifying detergent may get them clean and bright, but does little, if anything, to remove and destroy the bacteria or other small organisms that can cause disease. To destroy these microorganisms, a germicidal soap or disinfectant must be used. These household products contain chemicals to kill bacteria and other disease-causing germs. Naturally, these special soaps and disinfectants must be safe for humans and be effective in both soft and hard water.

The most commonly used disinfectants contain chlorine or a compound of chlorine, usually sodium hypochlorite or calcium hypochlorite. These chemicals have proven their effectiveness in destroying disease-causing bacteria. The addition of chlorine or one of its hypochlorite compounds to water supplies has practically eliminated many diseases, including typhoid fever, dysentery, and cholera. These diseases, transmitted by bacteria in water, killed many thousands of people each year before the chlorination of water supplies became a common practice in many areas of the world. In addition to being placed in water supplies, chlorine is often added to swimming pools to prevent bacterial contamination.

Disease-causing germs can also be destroyed by several other chemicals, including carbolic acid, potassium permanganate, ozone, iodine, and hydrogen peroxide. The effectiveness of these compounds varies; some, such as carbolic acid, have long-lasting effects while others, such as hydrogen peroxide, must be used repeatedly. In the following investigation, you will have the opportunity to examine the effectiveness of different disinfectants by testing their ability to control the growth of bacteria.

Help Stamp Out Bacteria

1. Obtain several sterile petri dishes containing nutrient agar. The agar acts as a surface on which the bacteria can grow; the nutrients in the agar serve as the food source needed for bacterial growth and reproduction.
2. Select an area in your home or school where you wish to investigate the extent of bacterial contamination. Remove the cover from all but one of the petri dishes and leave the agar exposed for five minutes. The one agar dish left unexposed will serve as a *control* for the experiment. A control is needed to make sure that only one factor or variable is responsible for the results of the experiment. For example, if bacteria appear in any of the exposed dishes but not in the unexposed one, then they must have been introduced when the agar plates were uncovered.
3. Replace the cover on one of the exposed petri dishes. To each of the others, add enough disinfectant solution to form a thin layer on the agar surface. If you are testing disinfecting aerosols, spray some into a beaker and wait until it liquefies. Replace the covers and incubate all the petri dishes, preferably at 37°C, for several days.
4. Observe the dishes daily. Compare them to those illustrated in Figure 8–4. Which ones show bacterial growth? Does one agar culture have significantly less bacterial growth? Which disinfectant was the least effective?

Chlorine—A Double Agent

Chlorine compounds can be useful not only as disinfectants but also as bleaching agents. Bleaching is a chemical reaction that removes the color from a substance. When used for laundering clothes, bleaches preserve the white color and remove stains. The bleaching process depends upon the chlorine reacting with the chemicals responsible for the spot or stain; once combined, the

Figure 8–4. The clear areas on these two petri dishes represent regions free of bacterial growth. The petri dish on the left had two drops of one liquid disinfectant added to two different locations, while that on the right had two drops of a different disinfectant brand. Which disinfectant is more effective?

colored material is changed into a colorless compound. The chemicals that caused the stain are still in the fabric—they just can't be seen because they've become colorless.

Care must be exercised when using bleaches containing chlorine compounds such as sodium hypochlorite. These chemicals are powerful bleaching agents that can damage the delicate fibers in silk and wool fabrics. Repeated use of these bleaches may also cause cotton fabrics to turn yellow. Adding a bluing agent to the wash may help in whitening these clothes. Since blue and yellow are complementary colors, they appear white when combined in the proper proportion.

A more important concern stems from a potentially dangerous reaction that can occur between chlorine bleaches and certain household cleaners. If a cleanser containing a base or acid is added to the bleach, poisonous chlorine gas is produced. To avoid a potentially hazardous situation, carefully read the directions and warning information on bleach containers. You can always use a nonchlorinated bleach to be perfectly safe. Although they are not as powerful, bleaches containing hydrogen peroxide or sulfur dioxide can change a colored substance into a colorless one. But will it be as white? To find out, try the following investigation. Since this experiment involves the production of a small amount of chlorine gas, be sure to carry out all the steps under a ventilating hood! If one is not available, omit adding the acid to the chlorinated bleach. Simply compare the whitening action of various brands of bleaches by soaking a small piece of colored cloth in the different solutions.

Bleach It Out!

1. Pour 25 ml of each bleach into separate 250-ml beakers. Add 100 ml of distilled water to each sample.
2. Place a small piece of colored cloth in each solution. You may

be interested in testing different fabrics to see how they bleach. Allow the samples to soak for 30 minutes, and record your observations.

3. Carefully add 20 ml of 1N hydrochloric acid to all of the chlorinated bleaches. Observe if any reaction occurs. Allow all the samples to soak overnight, and the next day record your observations. Are nonchlorinated bleaches as effective if allowed to stand overnight?

Topics for Further Investigation

Driers often cause "static cling," causing delicate fabrics to stick together. Fabric softeners added to the wash cycle or specially treated cloths added to the drier can help reduce this static cling. Can you design an experiment that can be conducted in a chemistry lab to test the effectiveness of these products? You may want to check with a physics teacher to see how to create the static electrical charge that causes fabrics to cling.

Because of their water-polluting effect, some phosphate-containing detergents were banned in many communities. However, many residents circumvented this ban by purchasing these detergents in neighboring communities where they were legally available. If you were the commissioner of the water supply in the community where these detergents were banned, what would you do?

Most manufacturers have their own combination of ingredients for the saponification process based on their own special formulas, especially in making liquid soap. Contact various manufacturers, requesting an explanation of the chemical reasons for including each of the ingredients listed on the label.

Obtain water samples from a swimming pool at various times throughout the year. Check for the extent of bacterial contamination. If your results fluctuate, can you suggest a reason for your observations? You may want to check the water at regular intervals between chlorinations. Does the chlorine have the same effect throughout this time period?

Investigate the chemical reasons why detergents, but not soaps, can lather in both hard and soft water.

Bleaches change the chemical structure of colored substances so that they turn white. Investigate the chemistry of color, especially dyeing processes responsible for changing one color into another.

Select a biodegradable compound found in a laundry room. Trace its history, beginning with its preparation by the manufacturer and ending with its decomposition by microorganisms. Explain the chemical action brought about by these decomposers.

*** The emulsifying action of soap and detergents depends upon the polarity of the charges present in their molecules. Containing both hydrophilic and hydrophobic ends, these emulsifiers are involved in ionic reactions, resulting in the dispersion of fats and grease in water. Investigate the chemistry of these ionic reactions. What factors influence the rate of these reactions?

*** Bleaches can function as either oxidizers or reducers. Compare the chemistry of these two processes. How do they compare in terms of efficiency, rate of action, and longevity? Write out the chemical equations, showing how both oxidizing and reducing bleaches work.

9 Fabrics from Fibers

Now that your clothes are clean, bright, and free of germs, let's take a closer look at their chemical structure. Fabrics used for making clothes are composed of fibers, long molecules shaped like fine threads. These fibers are twisted around one another, forming thicker strands that can be woven into a shirt, blouse, blue jeans, or any other article of clothing. Depending upon their origin, fibers are classified as either natural or synthetic. Natural fibers are obtained from animals and plants and have been used for making clothes for thousands of years. For example, early hunters stripped the hair fibers from animals that were killed primarily for food. These hairs were then woven into body coverings to provide warmth.

As early civilizations turned to agriculture as a means of providing a more dependable food source, people began to use fibers from plants for weaving fabrics. Since these plant fibers resulted in softer and more comfortable fabrics, their use rapidly increased as people recognized their value for making clothes. In fact, fine clothing produced from these fibers became a mark of advanced civilizations. The ancient Egyptians wove cloth fabric from natural fibers, especially flax, a long-stemmed plant yielding fine, light-colored fibers commonly known as linen. Early Europeans discovered that warm clothing could be made from certain animal fibers, especially wool.

Today, the textile industry utilizes four main natural fibers for making clothes—silk and wool from animals, and cotton and linen from plants. Although these natural fibers were perfectly acceptable to past generations, they are often unsuitable for weaving fabrics that meet the demands and expectations of the modern

121

consumer. If you examine either wool or silk fibers under a micro-scope, you'll recognize one reason why textile manufacturers ex-plored ways to synthesize new types of fibers. Wool and silk fibers have uneven surfaces, causing the fabrics to wrinkle easily. In ad-dition, their durability, resistance to staining, and resilience to repeated cleaning in washing machines left much to be desired. Moreover, increased consumer demand for fabrics exceeded the supply available from natural resources, making it even more ur-gent for the textile manufacturers to find ways of synthesizing new fibers. Today, nearly two-thirds of the world's clothing is made from synthetic fibers, artificially produced by special chemical processes.

The synthesis of these fibers begins by chemically treating certain raw materials, especially petroleum and wood products. When synthetic fibers are produced, they sometimes consist of molecules that are twisted, randomly arranged, and rough. Forc-ing the fibers through very narrow metal openings produces smooth and evenly shaped threads. Subsequent treatment with various chemicals can change the luster, silkiness, durability, strength, elasticity, and many other features that make the syn-thetic fibers more appealing and useful to consumers. In fact, the manufacturing process used in the textile industry can be consid-ered chemical magic since certain synthetic fabrics produced are difficult, if not impossible, to distinguish from natural ones. Among such synthetic fabrics are "leathers" and "suedes" that have never been near an animal—other than a human. In addition, some of the silkiest fabrics are not silk at all, white "cottony-soft" clothes may not contain any cotton fibers!

Since visual examination of a fabric may not reveal its true identity, only a chemical analysis can show whether the fiber is natural or synthetic. In the following investigation, you will con-duct several chemical tests on both natural and synthetic fibers. Based on your observations, you can then analyze a piece of cloth-ing to determine if it has been woven from natural or synthetic fibers.

Is the Fiber Real or Fake?

1. While holding a few cotton fibers in a pair of tongs, ignite them with a match. Note the odor produced, the amount of residue left, and the rate of burning. Repeat this procedure with several samples of both natural and synthetic fibers. Natural fibers from plants emit an odor of burning paper, while those from animals produce an odor of burning feathers. All fibers will

leave a black, charred residue indicating the presence of the element carbon. However, synthetic fibers will form small, round beads when burned. Check any unidentified fabric samples for the presence of natural and synthetic fibers. Are the odors produced when heated, the types of residue remaining, and the rate of burning typical of natural or synthetic fibers?

2. Place a small sample of each fiber in a test tube. Moisten a piece of red litmus paper with distilled water. Using forceps, place the strip in the opening of the test tube and secure it with a rubber stopper as shown in Figure 9–1. Be sure to use a one-hole stopper to avoid any buildup of pressure that could blow out the stopper. Carefully heat the test tube. Note whether the red litmus paper changes color. When heated, animal fibers containing protein will give off nitrogen. This nitrogen will react with the water on the litmus paper, producing the compound ammonium hydroxide. Like all bases, ammonium hydroxide will turn red litmus paper blue. Test any unidentified fabrics for the presence of proteins.

3. Repeat the procedure in step 2, substituting filter paper moistened with 10 percent lead acetate solution for the red litmus paper. The formation of a dark brown or black color on the paper indicates the presence of sulfur, an element present in protein. The sulfur reacts with the lead acetate to produce black lead sulfide. Be especially careful while heating the test tube, since any black soot produced by overheating will give a false positive test with the lead acetate paper. Which of the unidentified fabrics contain animal fibers as evidenced by positive tests for proteins?

Consumer Concerns

In the preceding investigation, you discovered that animal fibers may be identified by the odor of burning feathers or the presence of proteins, while plant fibers could be recognized by the odor of burning paper. Synthetic fibers, on the other hand, neither produce these characteristic odors when burned nor given positive tests for proteins. Although these differences were helpful in distinguishing between natural and synthetic fibers, obviously they are not important to a consumer, like yourself, who is concerned about qualities such as their durability, resistance to wrinkling, elasticity, color, and strength.

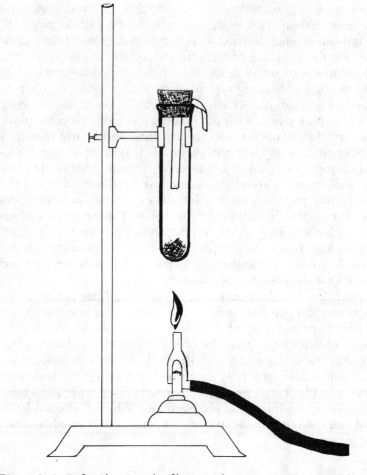

Figure 9–1. Before heating the fibers, make sure to use a one-hole rubber stopper to secure the paper strip so that it hangs into the test tube. The hole will allow gas to escape and prevent a buildup of pressure that could cause the test tube to explode.

The particular quality of a fabric depends on the chemical nature and physical arrangement of its fibers. If the molecules in the fiber are held close together by strong chemical bonds, then the fabric may not show much elasticity and may be difficult to weave. This type of fabric may also become brittle with age and break easily if subjected to constant wear. However, this fabric may be resistant to wrinkling and wash and dry easily. On the other hand, if the molecules in the fiber are more flexible, then the fabric could stretch and be useful for clothing, such as athletic outfits and sweaters. If the fibers are too elastic, however, the fabric will stretch so much that it sags. Have you ever had a favorite sweater that became too baggy before you outgrew it?

Is there a fabric that will stretch but readily return to its original shape, be strong but not break apart when pulled, and repel water but absorb dyes? To find out, conduct the following exercise to examine some of the qualities of various fibers. Obtain some samples of both natural and synthetic fibers. The latter might include nylon, Orlon, Dacron, Banlon, and rayon.

Give the Fiber the Stretch Test

1. Tease apart the fibers from each fabric and examine them under the low power of a microscope. Sketch each type. Can you identify the fabrics illustrated in Figure 9–2 by comparing them to those you've examined? Under each sketch, include some descriptive comments. Are the fibers organized in some way? Are they close together or are air spaces visible between them? Do they have smooth or irregular surfaces? Are they straight or coiled?
2. Pull hard on several fibers of each type. Rate each type of fiber on a scale from 1 to 5, with 5 representing the most difficult one to break. Repeat this test after wetting each type of fiber with water. Are the results different when the fibers are wet?
3. Gently stretch several fibers of each type and determine their ability to return to their original shape. Again, rate each fiber from 1 to 5, with 5 representing a fiber that quickly resumes its original shape.
4. While holding a few fibers from each fabric in a forceps, ignite them with a match. Rate each one from 1 to 5, with 5 representing the least combustible fiber. Construct a table indicating each type of fiber and its rating with respect to strength, elasticity, and combustibility.
5. Wet each fabric with several drops of water and observe if the fibers absorb or repel the water. Record which one dries the fastest.
6. Cut several fibers from each fabric into 10-centimeter lengths. Place a few fibers of each type on separate evaporating dishes. Label each dish, indicating the type of fiber, and place them in an incubating oven for 30 minutes at 60°C. Remove the dishes from the oven and measure the length of the fibers in each dish. Calculate how many centimeters each fiber has shrunk, divide by 10, and record the percentage of shrinkage for each fiber.

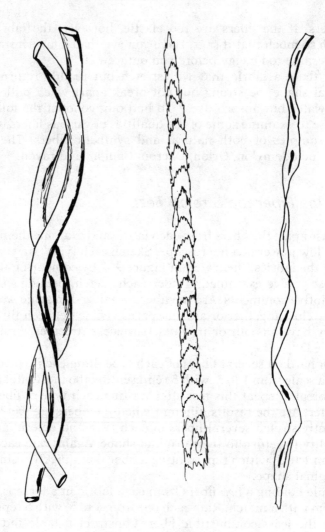

Figure 9–2. Can you identify these fibers after comparing them with ones examined under a microscope?

Based on your results, which fabric is best suited for making each of the following articles: parachutes, raincoats, flame-retardant pajamas, scuba-diving suits, and wash-and-wear shirts? Why are stockings made of nylon? Why are summer clothes made from cotton, and winter ones from wool? What chemical and physical properties should the fabric used for making astronauts' suits have?

Read the Label Carefully

If you examine the labels on some of your clothes, you'll probably discover specific instructions for washing or cleaning them: Don't use hot water, dry clean only, avoid adding bleaches, do not soak in strong solutions, etc. Since the molecules in the fibers can react with certain chemicals, it's a good idea to follow the manufacturer's suggested washing procedure. Some fabrics may be weakened and even dissolved by chemicals, while others may become spotted when mixed with certain compounds. In the following experiment, you will have the opportunity to test the effects of chemicals on various fabrics. Your observations may prove useful next time you wash that expensive wool sweater or your favorite cotton shirt.

1. Place a small piece of fabric in seven test tubes labeled A through G. Place 10 ml of each of the following chemicals in the test tubes. Be careful when pouring these solutions because some can cause skin irritation or burns. If feasible, each group of students can investigate a different type of fiber. A table summarizing the results can be constructed upon conclusion of the analyses.

      test tube A—3M hydrochloric acid
      test tube B—3M sodium hydroxide
      test tube C—chlorine bleach
      test tube D—dry-cleaning solvent
      test tube E—liquid soap solution
      test tube F—liquid detergent solution
      test tube G—acetone

2. Stopper each test tube and shake well for one minute. Allow the fabrics and solutions to stand for 20 minutes with occasional mixing.
3. Remove the fabrics with a glass stirring rod and blot any excess liquid with paper towels. Observe the fabric for any change in color, texture, or appearance. Which solutions were most damaging to the fabric?
4. Thoroughly wash the fabric samples with water. After blotting them dry with paper towels, check for any changes in elasticity or strength by pulling them. Which solutions caused the greatest change?

Making Your Own Fabric

If nylon was one of the fibers analyzed in the preceding experiment, you probably discovered that it is superior to the other fabrics in several respects. Made from coal-tar derivatives, nylon is resistant to bleaches and bases, does not react with organic solvents since it is chemically inert, and washes and dries more easily than any other fabric. Because of these properties, nylon has become a popular fiber for making a wide variety of consumer goods, including carpets, lingerie, hosiery, tents, toothbrushes, parachutes, tennis racquets, and blouses.

The properties of nylon stem from its chemical structure. One of the chemical compounds used in synthesizing nylon can react at

Figure 9–3. By slowly rotating the beaker, see how long a fiber can be extracted from the point where the two liquids meet.

both ends, making it possible to string many molecules together into a long chain known as a *polymer*. The following exercise involves the formation of a nylon polymer by layering one solution on top of a different one.

1. Prepare a 0.5M hexamethylene diamine solution by dissolving hexamethylene diamine in 0.5M sodium hydroxide. Pour 25 ml of this solution into a clean, dry beaker.

2. Carefully pour 25 ml of 0.25M sebacyl chloride or 0.25M adipyl chloride solution prepared in hexane down the side of the beaker. Tilt the beaker so that one solution will be layered on top of the other. Since the two solutions are immiscible, like oil and water, they will not mix but form two layers. Closely examine the junction of the two layers, and record your observations.

3. With a pair of tweezers or a paper clip bent in the shape of a hook, grasp the film between the two layers and pull upward. Wrap the threads around a small beaker, as shown in Figure 9–3. Slowly rotate the beaker and see how long a thread can be extracted from the two solutions.

4. Carefully wash the nylon threads and blot them dry with paper towels. Compare the strength, elasticity, and chemical properties of the nylon you synthesized with commercial ones you may have tested.

5. After the fibers have dried, ignite them with a match. Record your observations.

Will the Fiber Burn?

As you noticed, nylon melts without producing a flame. Concerned about fire safety, the United States government passed a law in 1953 to prohibit the sale of clothing that ignited too easily or burned too quickly. After chemists had developed procedures for chemically treating fabrics to retard flames, a law requiring flame retardants in childrens' pajamas was enacted in 1974. The chemicals used in this process had to be durable, resistant to organic solvents, and capable of remaining bonded with the fabric after repeated washings. The following experiment allows you to explore the flame-retarding properties of several chemicals.

1. Obtain several pieces of white cotton cloth. Hold one of the pieces with tongs and ignite it over a Bunsen burner. If possible, conduct this procedure under a ventilating hood. In any case,

make sure that there are no flammable materials in the area. Record your observations.

2. Soak two pieces of cotton cloth for five minutes in a solution containing a flame-retarding chemical. Among those you can try are saturated solutions of sodium tetraborate and sodium silicate. Remove one of the cloths, allow it to dry, and then ignite it over a Bunsen burner. Did any of the solutions make the cloth flame-resistant, flameproof, or both?

3. Remove the second piece of cotton cloth from the solution and rinse it thoroughly with running water. After allowing the cloth to dry, ignite the fabric over a Bunsen burner to determine how much of the flame-retarding chemical is left. Do any of the chemicals remain bonded to the fabric after a thorough washing?

4. By the way, an effective flame-retardant solution to spray on a Christmas tree can be prepared by dissolving 60 gm of ammonium sulfate, 30 gm of boric acid, and 7 gm of borax in a liter of water. This solution can also be poured into the support stand.

Fabrics for the Good Old Summertime

Cotton fabrics, similar to those examined in the flame-retarding analysis, are often used to make summer apparel. Several chemical properties of cotton, including its ability to absorb moisture, its fineness, light weight, and whiteness, are ideal for hot weather. Cotton, like linen, is made from the fibrous, woody parts of a plant. Like nylon, cotton is also a polymer. Whereas nylon is a synthetic fabric made from coal-tar derivatives, cotton is a natural plant product consisting of long chains of glucose molecules strung together to make a polymer known as *cellulose* (see Figure 9–4).

In fact, cotton is almost all cellulose. Although cellulose makes cotton suitable for summer apparel, it lacks certain chemical properties that would make the fabric even more appealing to consumers. Recognizing that the use of cotton for making clothes could be greatly expanded, textile manufacturers developed chemical processes to make the cellulose fibers more resistant to fire, less likely to shrink, and easier to dye. Moreover, the use of ether to remove the natural wax found on cellulose greatly increases the absorbency of cotton, making it valuable for medical purposes.

Because of both its natural properties and those produced by

Figure 9–4. Glucose units are connected to one another to form cellulose. This process continues until the large polymer is completed.

specialized chemical processes, cotton has become the major fiber used in the textile industry. Much of cotton's popularity stems from its ability to be dyed. Recognizing that cotton would be more appealing if it were colorful, people have used dyeing processes for centuries. The first dyes were extracted from colorful plants and animal skins. Today, most dyes used by the textile industry are synthetically produced.

The following exercise will explore the two major ways by which fibers can by dyed. The first is a direct method where the dye chemicals are attached to the fibers in the fabric. The second method is an indirect process; the fabric is first treated with a chemical substance known as a *mordant*. Mordants are compounds that bond to the fabric and do not wash out after repeated launderings. The dye is then combined with the mordant. To find out how colorful you can make white cotton cloth, try both methods!

Color Your Fabric

1. The sizing, the material found in the pores of fibers, must first be removed so that the dye can reach the fabric. To accomplish this, place several pieces of white cotton cloth in a 1-liter beaker. Add enough 0.25M sodium carbonate solution to cover the cloths and boil for two minutes (see Figure 9–5). After allowing the solution to cool, remove the cotton cloths, rinse thoroughly in running water, and wring them dry. With the sizing removed, the dye can now be added.
2. Dissolve 2 gm of a dye in 1 liter of distilled water. Dyes you can use include methyl orange, Malachite green, crystal violet, Congo red, and fuchsin. If available, commercially prepared dye solutions can be used. For direct dyeing, immerse two pieces of cotton cloth in 300 ml of dye solution placed in a 1-liter beaker. Make sure the solution covers the fabric and boil for five minutes. Remove the cloths, rinse with water, wring them out, and hang up to dry. Save the dye solution for step 5.
3. To add a mordant, begin by pouring 300 ml of 0.5M aluminum sulfate solution into a 1-liter beaker. Immerse two cotton cloths in this solution and boil for three minutes. Remove the cloths and rinse them with water.

Figure 9–5. Indirect dyeing first involves heating the fabric in a solution containing a mordant.

4. Pour 200 ml of 0.5M sodium hydroxide into another 1-liter beaker. Place the two cloths that had been boiled in the aluminum sulfate solution into the sodium hydroxide solution and boil for one minute. A mordant has now been added to the cotton fibers. Remove the cloths, wash with water, and wring dry.
5. Place the cloths from step 4 in the dye solution saved from step 2 and boil for five minutes. Remove the cloths, rinse with water, wring them out, and hang them up to dry.
6. Is there any difference in color between the cloths that were dyed directly and those dyed after adding a mordant?

Will the Fabric Bleed?

A fabric that retains the dye after washing is colorfast and said not to "bleed." Not only must the dye withstand repeated launderings, but it also must resist fading by light, perspiration, and aging. Can you design an experiment to test the colorfast quality of each of the fabrics you dyed in the preceding investigation? Are fabrics that have been dyed directly more colorfast than those first treated with a mordant? Observe whether the color changes, or if it bleeds with no visible change in color.

If you plan to test the effect of perspiration on dyes, be aware that "perspiration" from different people can vary from acidic to basic. Consequently, you will have to prepare both acidic and basic perspiration solutions. The fabric should be thoroughly wet with the solution, wrung out, and then placed in an oven at 37°C for 6 hours. Why is 37°C the recommended temperature?

Topics for Further Investigation

Fibers can be prepared from minerals. Among such fibers are glass, asbestos, and fiberglass. Investigate the manufacturing process used in the making of mineral fibers.

When analyzing foods for the presence of preservatives, you prepared an ester. Dacron is a polyester, consisting of a polymer of ester units. Can you design an experiment to synthesize a polyester fiber?

Coffee, tobacco, grape juice, mustard, and lipstick stains may be difficult to remove from fabrics. Laundry bleaches containing chlorine are usually effective in removing such stains by use of a chemical reaction known as *oxidation*. Examine the chemical basis of this stain-removing process.

Compare the preparation of both natural and artificial silk fibers. The first synthetic fiber was artificial silk.

Design an experiment to determine which fiber is easiest to dye. Which fiber is the least likely to bleed? Can you explain the chemical basis for your observations?

How do various temperatures of the water in washing machines and irons affect synthetic fabrics? Can you design a laboratory procedure to check these effects?

Orlon, a synthetic fiber, is an acrylic fiber. Investigate the chemical nature of acrylics. Identify other synthetic products chemically similar to Orlon.

*** Invent a fiber that would be suitable for specialized purposes such as deep-sea diving, space travel, or fire fighting.

*** Explore the feasibility of using the technology of genetic engineering to produce bacterial strains that will synthesize a particular fiber. Can the appropriate genes from the silkworm be isolated and recombined with a bacterial chromosome?

CHEMISTRY IN THE GARAGE

10   *Petroleum Products*

Animals that once roamed the earth now fuel your car! After these prehistoric animals died, their fossilized remains decayed, forming layers on the earth's surface. Over millions of years, these layers were covered with other deposits and became buried deep in the earth. Intense pressure and heat gradually changed some of these layers into a variety of organic compounds, including one important group known as petroleum. Formed from the fossils of prehistoric organisms, petroleum is a mixture of compounds composed primarily of the elements carbon and hydrogen.

Compounds made entirely from carbon and hydrogen are called *hydrocarbons*. Although only two elements are present, hydrocarbons come in a wide variety of sizes and shapes. The simplest hydrocarbon, methane, consists of one carbon atom bonded to four hydrogen atoms, as illustrated in the following structural formula:

If you skipped the section dealing with the chemical structure of organic compounds in Chapter 3, all you need to understand about a structural formula is that each atom is represented by a letter—in this case, C for carbon and H for hydrogen. The atoms are connected to one another by bonds, illustrated as lines in a structural formula. Because of its chemical structure, carbon can bond with other carbon atoms, forming long chains constituting a carbon backbone. For example, ethane contains two carbons bonded to six hydrogen atoms. Knowing that carbon must form four bonds while hydrogen forms only one, can you draw the structural formula for ethane? The process of bonding one carbon atom to another can continue, producing hydrocarbons such as butane (four carbons), hexane (six), and octane (eight). Try your hand at drawing the structural formulas for these compounds.

Not only can carbon atoms form long chains, they can also bond to make cyclic or ringlike compounds, as illustrated in the following structural formula for benzene:

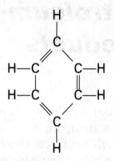

If you closely examine this structural formula, you'll notice that two lines connect some of the carbon atoms, indicating the presence of *double bonds*. When considering the different ways of arranging carbon and hydrogen atoms using long chains, ring structures, and double or even triple bonds, you can imagine the great variety of structures possible for hydrocarbons. Many of these hydrocarbons are used as fuels, including butane, propane, gasoline, kerosene, diesel fuel, and home heating oil.

All these fuels are derived from petroleum—that thick, dark-brown liquid that supplies much of the world's energy, including that needed for running our cars. However, petroleum, as it comes out from the earth in its natural state, cannot be used as a fuel. When pumped out of the ground where it has been for millions of years, petroleum contains a mixture of many different hydrocarbons. Before petroleum can be made useful as an energy source, the various hydrocarbons must be separated by a refinery process known as *fractional distillation*.

This refining process depends upon the differences in the boiling points of the various liquid hydrocarbons present in petroleum. During fractional distillation, a liquid mixture is heated; the liquid with the lowest boiling point will vaporize into a gas first.

Cooling this gas in a process known as *condensation* causes it to form into a liquid again so that it can be collected and isolated from those hydrocarbons that have not reached their boiling point. As the temperature is increased, hydrocarbons with successively higher boiling points are vaporized, condensed, and isolated. The following experiment demonstrates how fractional distillation operates.

Separating the Parts in Alcohol

1. Place 50 ml of rubbing alcohol and several boiling chips in a 250-ml flask. Insert a thermometer into a two-hole rubber stopper that fits securely into the flask. Do not force the thermometer through the hole. If you have any difficulty, moisten the bulb with glycerin and slowly twist the thermometer to work it through the stopper. Position the thermometer so that it rests 2 cm from the bottom of the flask; it must be immersed in the rubbing alcohol.

2. Bend a piece of glass tubing to form a right angle. Insert one end into the rubber stopper and connect a piece of rubber tubing, at least 1 m long, to the other end. Stopper the flask, making sure all connections are tight. If you do not have a Liebig condenser, connect a short piece of straight glass tubing to the other end of the rubber hose and place it in a graduated cylinder immersed in an ice bath (see Figure 10–1).

3. Making sure that the graduated cylinder and ice bath are as far away as possible from the flask, heat the alcohol on a hot plate. Be careful, because the vapors from the rubbing alcohol are flammable.

4. Record the temperature when the first liquid vaporizes and condenses in the cylinder. The temperature will stabilize until all this liquid has vaporized. When the temperature again begins to rise, replace the graduated cylinder in the ice bath with another one.

5. Continue heating until another liquid vaporizes and condenses in the cylinder. Record this temperature. Stop heating the alcohol while there is still a small amount remaining in the flask.

6. Can you identify the first liquid you collected? What is its boiling point? What is the second liquid and its boiling point?

7. Can you prepare a mixture of several liquids and then successfully separate each one by fractional distillation?

Figure 10–1. Fractional distillation involves heating a solution to separate the various components, which can be individually collected in a graduated cylinder immersed in an ice bath.

Take the Hydrocarbons in Order

During fractional distillation, the lightest hydrocarbons, or those with the fewest number of carbon and hydrogen atoms, are the first to be separated. In fact, these compounds are so light that they often bubble out of the petroleum as soon as the layer is exposed by drilling. Occasionally, they separate from the heavier hydrocarbons while underground, forming pockets of natural gases. These gases can be tapped and sent in pipelines throughout the country for use in cooking, heating, and industrial purposes.

Next comes gasoline, which is mostly a mixture of seven carbon (heptane) and eight carbon (octane) compounds. Gasoline vapors are extremely volatile and are ignited easily by sparks such as those produced in a car's engine. The chemical energy in the gasoline can then be released to produce the mechanical energy needed to run the car. However, igniting gasoline is a violent chemical reaction, producing engine "knock." Lowering both the temperature and pressure at which the gasoline can be ignited reduces the knocking sound, but it also decreases the efficiency of the gasoline by yielding fewer miles per gallon.

The efficiency of gasoline is listed as the octane number. The higher the rating, the less knock is produced when the gasoline is ignited. An octane number of 0 is given to heptane and 100 to octane. A gasoline sold at a service station with an octane rating of 90 has the same antiknock quality as a mixture of 90 percent octane and 10 percent heptane. Using a gasoline with a sufficiently high octane number will prevent engine knocking. Buying a gasoline with an octane number higher than that recommended for your car is a waste of money, because the gasoline will not improve the engine's performance.

Octane ratings can be improved by various methods. The addition of lead-containing compounds to gasoline was a common process used to reduce knocking. Since the exhaust emitted from burning gasoline containing lead causes air pollution, car manufacturers are now required by law to design engines that use unleaded gasoline. To improve the octane rating of unleaded gasolines, refineries use a process known as *cracking*, a method by which larger compounds are changed into smaller ones. Breaking down the heavier hydrocarbons in gasoline into lighter ones that vaporize more easily when ignited increases the efficiency of the gasoline.

After gasoline, the next fuel isolated by fractional distillation is kerosene. With higher temperatures, jet fuel and then diesel oil are isolated. As illustrated in Figure 10–2, the last hydrocarbons separated out of the petroleum mixture are the lubricating oils and greases, each one containing sixteen or more carbon atoms as part of its structure. These heavier hydrocarbons are used to reduce the friction produced when the moving parts of machinery rub against one another. Lubricating oils and greases allow such parts to slide easily rather than grinding hard against each other. Without motor oil, the parts of your car's engine would grind against one another, eventually stick together, and finally stop moving. Your car would come to a grinding halt!

Motor oils are fluids with a high viscosity. The higher the viscosity, the more slowly the liquid flows. The high viscosity is an important property of lubricating oils since they stick to the metal surfaces, keeping them slippery and reducing friction. The viscosity can be determined by measuring the flow of the liquid through a narrow opening at a specified temperature.

The viscosity of motor oils is listed on the can. The Society of Automotive Engineers (SAE) developed the system of listing the viscosity as an SAE number—for example, SAE 20 or SAE 10W–40. If the viscosity is measured at 99°C, then a single SAE number is assigned: SAE 10, SAE 30, etc. Those oils measured at both 99°C

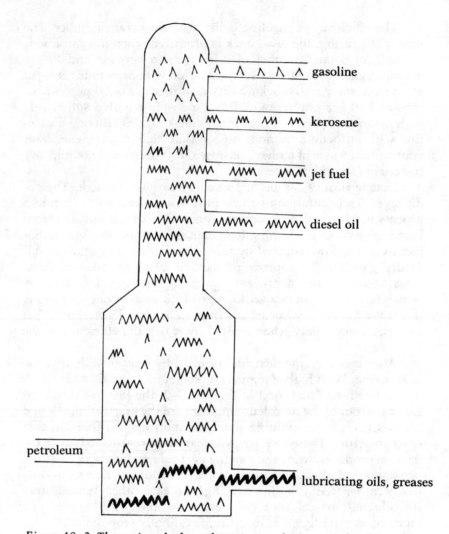

Figure 10–2. The various hydrocarbons in petroleum are isolated in a fractionating tower on the basis of their weight and size.

and −18°C are given another number followed by a W: SAE 10W–40. The 10W indicates the viscosity measured at −18°C, while the 40 refers to the viscosity determined at 99°C. Obviously, the viscosity changes with temperature. In the case of the SAE 10W–40 motor oil, the viscosity increases by a factor of four when the temperature is raised from −18°C to 99°C.

Can you explain why a motor oil should be *less* viscous in cold weather and *more* viscous in warmer weather? The following experiment investigating the viscosity of motor oils with different SAE numbers may provide some clues. Multigrade oils, or those whose viscosity has been measured at both temperatures, should be included in your study.

How Thick Can Oil Get?

1. Place 10 ml of water in a buret. While watching the second hand on a clock or watch, open the stopcock and record the time it takes for the 10 ml of water to empty into a beaker. Repeat this procedure three times and calculate the average number of seconds.
2. Place 10 ml of motor oil in the buret and repeat three time trials as you did for water. Time the flow of as many different motor oils as possible. Thoroughly rinse the buret several times with soap and warm water before testing each brand. Rinse the buret with tap water to remove any remaining soap. Compile a table indicating the brand name, SAE number(s), and average time flow based on three trials.
3. If a buret is unavailable, you can measure the viscosity as you did for hair shampoos. Place 100 ml of water in a graduated cylinder and measure the time it takes for a small glass or steel bead to fall to the bottom. Determine the average time for three tests. Use this procedure to measure the viscosity of each motor oil, again keeping in mind that the graduated cylinder must be thoroughly washed before testing each brand.
4. Calculate the relative viscosity for each motor oil. The relative viscosity refers to how much more slowly the motor oil flows than the water and is determined by using the following equation:

$$\text{relative viscosity} = \frac{\text{average time (motor oil)}}{\text{average time (water)}}$$

5. Repeat the viscosity measurements after varying the temperature of the motor oils. Be very careful when heating the oil, since hot oil can cause serious skin burns. How does the relative viscosity of each oil change after cooling? after heating? Graphing your results may provide a clearer demonstration of the relationship between viscosity and temperature. Can you now explain why a motor oil for winter weather should flow more easily—have a viscosity closer to that of water? Why should motor oils for summer use have a higher viscosity? What would

the SAE number be for a winter-use motor oil? a summer-use motor oil? What is the advantage of a multigrade motor oil?

Oil in Winter

As the temperature drops during the colder months, you have to be concerned not only about the viscosity of motor oil but also about the level of antifreeze in the radiator. Water, functioning as a coolant, prevents the engine temperature from rising as a result of the tremendous heat released upon combustion of the gasoline. Although water is an effective coolant, it has the disadvantage of freezing at temperatures normally encountered during winter. To prevent the water in the radiator from freezing, another liquid, usually ethylene glycol, is added. Ethylene glycol functions as an antifreeze by lowering the temperature at which water freezes.

The effect of adding ethylene glycol is the same as that of dissolving a solute in water—they both lower the freezing point of water. But does varying the amount of solute added to water produce different effects on the freezing point, or will the water freeze at the same temperature regardless of how much solute is added? Can you design an experiment to answer this question? If you plan to use an antifreeze, vary the proportion of the ethylene glycol–water mixture. If you prefer, dissolve varying amounts of sugar in different water samples. To obtain the low temperatures needed to freeze these solutions, recall the use of the ice–salt mixture to determine the freezing point of water in Chapter 1. Placing the various solutions in a freezer would also accelerate the time required for them to freeze. Be sure to allow the solutions to thaw if the thermometer gets frozen in the solution!

Can ethylene glycol help your car's engine in summer? To find out, test the effect of adding ethylene glycol or a solute on the boiling point of water. Does the addition of a specific amount of ethylene glycol or solute have a greater effect on the boiling point or on the freezing point?

Can't Get Started? Check Your Battery

Even though your car's tank may have been filled with gas and had enough water to prevent overheating on a hot summer's day, you may not have been able to go anywhere because of a dead battery. The battery in your car supplies an electric current to start the engine. Chemicals with certain properties are required to produce this current by generating a flow of *electrons*. An electron, which is part of an atom, is a negatively charged particle that can be transferred from one atom to another. Metals are very efficient for initiating and conducting an electric current because their atoms easily and rapidly transfer electrons from one metal atom to another.

The battery in your car contains a group of lead plates immersed in a water solution of sulfuric acid, which acts as an *electrolyte*. An electrolyte is any substance that allows the transfer of electrons between atoms. An electric current is generated when electrons flow through the sulfuric acid solution from one lead plate to another (see Figure 10–3). If the electrons always flowed in the same direction, the battery would soon lose its charge. An alternator reverses the flow, restoring the battery to its original condition by causing the electrons to move in the opposite direction.

In the following investigation, you can examine the conditions needed to make a battery by establishing an electric current through a lemon or an orange! Since sulfuric acid causes skin burns, you can use the weak citric acid found in these fruits to serve as an electrolyte.

Can You Charge a Lemon?

1. With a sharp knife or razor blade, make two parallel slits in a lemon or an orange. Insert a zinc strip in one slit and a copper penny in the other, making sure that half of each piece of metal extends from the surface of the fruit.
2. Connect an alligator clip to the zinc metal and attach the wire to the negative terminal of a galvanometer. Connect the penny to the positive terminal, using another wire and alligator clip as pictured in Figure 10–4.
3. Record what happens to the needle on the galvanometer. Explain your observations. Keeping in mind that electrons carry a negative charge, which metal is losing electrons and which one is gaining electrons?
4. Experiment with different metals to determine which pair is most effective in establishing an electric current. Try different citrus fruits to see if they vary in their capacity to function as electrolytes. Based on your observations, can you explain why a battery dies when there is no water surrounding the lead plates?

electron flow

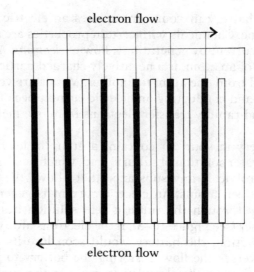

electron flow

Figure 10–3. Electron flow between lead plates submerged in sulfuric acid provides the current in a car battery. The alternator reverses the flow to reestablish the original conditions.

5. You may be interested in making a more powerful battery by wrapping aluminum foil around the outside of a large beaker and connecting it to the negative terminal of a voltmeter. Attach the positive end to another metal such as copper or lead. Fill the beaker with 0.1M hydrochloric acid and then immerse the copper or lead strip in the acid. Record how many volts of electricity are generated. What happens if you substitute 0.1M sulfuric acid in place of the hydrochloric acid?

Electron Flow Between Iron and Copper

Place an iron nail in a 0.1M copper sulfate solution. After ten minutes, remove the nail and describe what happened. The change in appearance of the nail was caused by a flow of electrons, similar to the one which occurred to create an electric current through the lemon. The iron in the nail lost electrons to the copper sulfate in solution. As the copper in this solution picked up these electrons, it turned into metallic copper, coating the surface of the nail. Whenever any substance loses electrons, in this case the iron, and another substance gains electrons, in this case the copper, an *oxida-*

Figure 10–4. With a galvanometer and some alligator clips, you can determine which fruits are the best conductors of electrical charges.

tion–reduction process has occurred. The substance losing electrons becomes oxidized, while the one gaining the electrons becomes reduced (see Figure 10–5).

An oxidation reaction is the cause of rust formation on a car. The iron present in the steel slowly reacts with the oxygen in the air. As the electrons are transferred from iron to oxygen, the compound iron oxide, commonly known as rust, forms. Is the oxygen being oxidized or reduced? To prevent or at least slow down the oxidation process, a car wax can be applied to the metal. As you may recall from your study of organic compounds, waxes are lipids that are solids at ordinary room temperature. A solvent is added to dissolve the waxes, either natural or synthetic, to liquefy the polish so that it can be more easily applied to the metallic surfaces of a car. After the polish is applied to the metal, it forms a hazy layer. A gentle buffing spreads out the polish, allowing the solvent to evaporate and leaving a thin wax layer that serves as a protective coating to inhibit the iron from oxidizing. The result is a shiny car.

A quick glance at the automotive section in a store would reveal a large number of products guaranteed to beautify and protect your car. But which one is best for inhibiting the oxidation reaction that results in rust? Can you design an experiment to

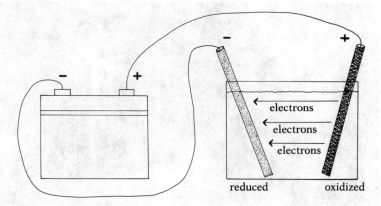

Figure 10–5. Any substance losing electrons is oxidized and develops a positive charge. The substance gaining electrons is reduced and becomes negatively charged.

evaluate commercial car waxes? In planning your procedure, you will need several pieces of metal from a car for testing each wax. Contact a salvage yard or junk dealer for the metal. Each wax should be tested for its ability to protect against temperature extremes, solutions simulating acid rain, prolonged exposure to sunlight, and repeated washings.

Unfortunately, no matter how effective the wax, a car will still rust because of the oxidation that begins on the inside surfaces. As the iron oxide accumulates, the metal rusts, beginning with the interior surfaces and working its way to the outside. To inhibit rust formation on the hidden surfaces, an undercoating can be applied. By the way, the chemicals in the undercoating are derived from petroleum!

 Petroleum Quiz

Having completed a study on the chemistry of petroleum products, you may be interested in typing the following program into the computer. This program randomly selects each of the questions in the DATA statements and asks if it is true or false. Examples of questions that can be asked are given in the DATA statements. You can either have someone type in different DATA statements to test your knowledge or enter these examples to challenge a friend or classmate. Of course, you can use this program with any DATA statements, including ones dealing with information obtained in previous chapters. If you type in different DATA statements, be sure each one is followed by a T (true) or F (false). Use the quotes in each DATA statement as indicated in the listing.

```
5 DIM Q$(10),A$(10)
10 PRINT "WHAT IS YOUR FIRST NAME";:INPUT NA$
20 FOR D = 1 TO 10
30 READ Q$(D),A$(D)
40 NEXT D
50 FOR D = 1 TO 10
60 A(D) = INT(10*RND(1)) + 1
70 IF D = 1 THEN 110
80 FOR B = 1 TO D − 1
90 IF A(D) = A(B) THEN 60
100 NEXT B
110 NEXT D
120 FOR D = 1 TO 10
130 PRINT Q$(A(D))
140 INPUT "IS THIS STATEMENT TRUE OR FALSE (T OR F)";D$
150 IF D$ = A$(A(D)) THEN GOTO 200
160 PRINT "THAT IS INCORRECT,";NA$;"."
170 PRINT
180 NEXT D
190 GOTO 240
200 PRINT "THAT IS CORRECT, ";NA$;"."
210 PRINT
220 C = C + 1
230 NEXT D
240 PRINT "YOU GOT ";C;" CORRECT ANSWERS, ";NA$;"."
300 DATA "IN WINTER, A CAR'S OIL SHOULD BE LESS VIS
      COUS","F"
310 DATA "FRACTIONAL DISTILLATION SEPARATES HYDRO
      CARBONS","T"
320 DATA "SOLUTES LOWER THE FREEZING POINT OF WA
      TER","T"
330 DATA "ELECTROLYTES ARE USED TO PREVENT OXIDA
      TION","F"
340 DATA "USE SAE 10 MOTOR OIL IN SUMMER","T"
350 DATA "KNOCKING IS CAUSED BY LEADED GASOLINE","F"
360 DATA "ANTIFREEZE RAISES THE FREEZING POINT OF WA
      TER","F"
```

370 DATA "REDUCTION IS THE LOSS OF ELECTRONS","F"
380 DATA "RUST IS CAUSED BY OXIDATION OF IRON","T"
390 DATA "ELECTROLYTES CAUSE A DEAD BATTERY","F"

Topics for Further Investigation

The burning of fuels is the most common source of energy in the world. With an increasing demand for energy, scientists are seeking alternative sources that are reasonably priced, reliable, abundant, nonpolluting, and efficient. Many believe that nuclear energy is the only answer. Do you agree? Defend your position with scientific arguments.

To reduce the air pollution caused by car exhausts, catalytic converters are required by law. Determine the chemical reactions undergone by the compounds emitted in exhausts and explain how the converters operate to reduce pollution.

Construct molecular models of the various fuels present in petroleum.

Pass an electric current through a beaker of water containing a small amount of dilute sulfuric acid. Design some way of collecting the gas produced at each of the electrodes placed in the water, and determine their chemical identity.

Since the fossil fuel supply is rapidly becoming depleted, scientists are investigating alternative energy sources for running cars, including rechargeable batteries and solar power. Can these sources be expected to provide energy for all the cars, trucks, buses, motorcycles, etc., currently powered by gasoline?

That shiny chrome bumper on your car was produced by a process known as electroplating. Investigate the chemical nature of this manufacturing process and explore its application in the production of a variety of materials.

** Recombinant DNA technology is being applied to develop a bacterial strain capable of swallowing up accidental spills from oil tankers. Investigate the feasibility of this approach by attempting to create a new bacterium that will metabolize the hydrocarbons in petroleum into harmless chemical compounds.

** Automotive engineers are continually searching for engines that would be more efficient and less polluting. Can you design a new catalytic cracking process to produce a gasoline that will give better mileage, knock less, and emit fewer pollutants upon combustion?

CHEMISTRY IN THE BACKYARD

11   *Outdoor Chemistry*

You're all set for a family barbecue when, suddenly, dark, rain-threatening clouds begin to appear and cover what was once a perfectly clear, blue sky. As soon as everyone sits down to enjoy those delicious-looking grilled hamburgers and hot dogs, the rain starts. While rushing to get the food inside before everything gets soaked, someone suggests that the next time a barbecue is planned, the family should listen to the radio for an updated weather report. Confident because of your knowledge of chemistry, you volunteer to inform everyone about the chance of rain the next time your family plans a barbecue—without even listening to a weather report! To find out how to be an accurate weather forecaster, you should know something about certain compounds that can be useful in analyzing the moisture content of the air. The following investigation involves the preparation of a chemical weather predictor.

What's the Forecast?

1. Prepare 100 ml of 0.1M cobalt chloride and 100 ml of 0.1M sodium chloride solutions. Mix the two solutions and pour into a 250-ml beaker.
2. Soak a piece of filter paper in the solution for several minutes. Record the color of the paper.
3. Remove the paper and allow it to dry overnight. What color is the paper the next day? As cobalt chloride dries, its color changes. Place the dry filter paper near an open window. An

151

increase in atmospheric moisture, indicating a potential rainstorm, will change the color of the filter paper. The addition of
the sodium chloride to the solution speeds the absorption of
moisture from the atmosphere by the cobalt chloride.

4. You can prepare a weather indicator to mount on the outside of
your house. Apply several coats of a saturated solution of cobalt
chloride to a metallic or plastic strip. Allow each coat to dry
thoroughly before applying the next one. When the strip is dry,
mount it to the outside of your house.

5. Have you ever wondered why the manufacturer put that small
packet inside the box of your new camera? Inside that packet
are special chemicals that readily absorb moisture. Known as
desiccants, these chemicals soak up the moisture to prevent the
mechanical parts from rusting. Check to see what chemicals are
used as desiccants by manufacturers of cameras, stereos, calcu-
lators, and watches.

6. If you wish to make a desiccator to dry a small object, begin by
covering the bottom of a wide-mouthed jar with either cobalt
chloride or calcium chloride. Bend the extending prongs of a
clay triangle to make support legs as shown in Figure 11–1.
Rest the clay triangle on the chloride compound. Place the object you wish to dry on top of the triangle and tightly seal the
jar. After a few days, note the color of the chloride compound.
Examine the object placed in the desiccator to see if any moisture has been removed.

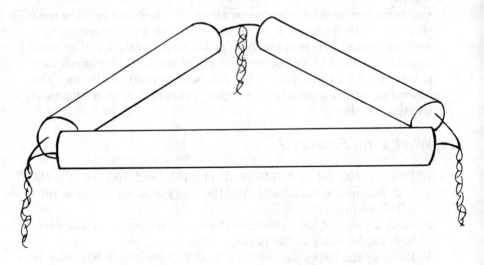

*Figure 11–1. Bending the wire extensions on a clay triangle produces a good
support stand for a desiccator.*

Rain Can Kill Your Plants

Rain may not only spoil your family barbecue but also damage plants, shrubs, bushes, and trees growing in your backyard. This potentially harmful effect of rain has recently come to scientists' attention. Known as *acid rain*, this problem is not confined to your backyard, but exists throughout the world, especially in and near heavily industrialized countries. As fossil fuels are burned to run factories, power cars, and provide electricity, gases emitted as by-products of these combustion reactions combine with water in the atmosphere. For example, carbon dioxide gas reacts with water to produce carbonic acid; sulfur dioxide gas combines with water to yield sulfuric acid; nitrogen oxide gases react with water to make nitric acid. Precipitation, either as snow, sleet, hail, or rain, deposits these acids back on earth, where they alter the pH of the soil and of the water in lakes, streams, ponds, and rivers.

Organisms living in lakes and ponds are extremely sensitive to changes in pH. Normally alkaline with a pH value of 8 or higher, these bodies of water contain natural buffers to prevent any fluctuation in the pH value. However, these buffers cannot always neutralize the acids collecting in lakes and ponds, especially from the spring runoff of winter snow. An acidic pH can inhibit the spawning of trout and bass, prevent the hatching of salamander eggs, and reduce the diversity of flora and fauna. In fact, if the pH reaches 4, no fish would remain; at pH 2, all insects would be killed. Obviously, the impact of increased acid accumulation in bodies of water can be devastating.

If you live near a lake or pond, you may find it interesting to check its pH value. As a long-term project, check the pH at regular intervals, noting the dates of rain, runoff from melting snow, or dumping of industrial wastes. Is there any one factor that has a significant effect on the pH? Check the area surrounding the lake or pond for rock formations. Rocks containing carbonate compounds are effective buffers, whereas noncarbonate rocks, such as granite, have no buffering capacity. Does the pH value remain more constant at a region near a carbonate rock formation? You can also take plankton samples to observe any changes in populations of microscopic organisms. If more than one lake or pond is nearby, conduct a comparative study to determine whether one becomes more acidified. In comparing your results, consider the characteristics of the land surrounding each body of water (see Figure 11–2). Are there forests, farmlands, factories, etc., present?

For those of you who do not live near a lake, pond, or any body of water, a number of experiments can be conducted under labora-

Figure 11–2. The chemical composition of a pond depends upon the nature of the surrounding area. Which pond probably would have a more acidic pH value?

tory conditions to observe the effects of acid precipitation on the environment. Using standard laboratory equipment, you can explore several aspects of this global problem, beginning with a chemical reaction to simulate the production of acid rain and concluding with an examination of the effects of acid rain on a small community that can be set up either in class or at home. As a result of completing these investigations, you will have a basic understanding of the problems posed by acid precipitation. With this knowledge, you will be in a better position to discuss this concern with local, state, and national politicians debating the issue of acid precipitation.

Bottle Some Acid Rain

1. Pour 250 ml of distilled water into a large jar. Add ten drops of one of the following indicators: bromocresol purple, chlorophenol red, bromophenol red, or bromothymol blue. Record the color of the distilled water and determine the pH value, using universal pH paper.
2. If you do not have a deflagrating spoon, bend a teaspoon so that it will hang on the rim of the jar, suspended over but not submerged in the distilled water, as pictured in Figure 11–3. Fill the spoon with powdered sulfur and heat over a Bunsen burner until the sulfur ignites.
3. Quickly and carefully place the spoon on the rim of the jar and cover tightly with aluminum foil. Allow the sulfur to burn out,

trapping the gas inside the jar. Remove the spoon but not the aluminum foil, keeping the amount of gas that escapes to a minimum. Swirl the jar to dissolve the gas in the distilled water. Record any change in the color of the indicator. Check the pH value of the solution.

4. Explain your observations. The burning sulfur reacts with oxygen in the air to produce sulfur dioxide, which then combines with the distilled water to give sulfuric acid. Check reference materials to determine the sources of atmospheric sulfur dioxide. Can you explain how sulfur dioxide can affect areas far removed from sites where it is first deposited in the atmosphere?

5. Repeat this investigation, first adding a small amount of calcium carbonate or limestone to the distilled water. Does the sulfur dioxide have the same effect? Why or why not?

What Is in Acid Rain?

In the preceding experiment, you prepared a solution comparable to acid rain. However, acid rain is not simply water with a pH value below 7. In fact, toxic metals, including lead, mercury, and cadmium, are often present in acid rain along with pollutants such as polychlorinated biphenyls (PCB) and certain hydrocarbons shown to be carcinogenic. If you consider the complex feeding relationships that can exist among different organisms, you'll recognize that the effects of acid precipitation can extend beyond the life found in a single pond or lake.

To appreciate the wide impact of acid precipitation, conduct the following investigation to explore some of the effects of acidic solutions on a variety of substances. Naturally, both organisms and nonliving materials may be exposed for prolonged periods of time to acidic solutions, especially in areas where bodies of water are located. Although your investigation will involve a relatively short exposure and omit the metals and carcinogenic pollutants, you will nonetheless have an opportunity to observe the types of substances and facets of life affected by acid precipitation.

1. Obtain a variety of materials; possibilities include marble, concrete, aluminum, zinc, copper, wood, glass, and paper. Place a small sample of each material in a small beaker.

2. Cover the sample with 3M sulfuric acid or 3M hydrochloric acid. Although the pH value of these solutions is less than that

Figure 11–3. The spoon is suspended over the distilled water. Once the sulfur is ignited, quickly position the spoon and cover the jar with aluminum foil to trap the gas.

of acid rain, you can still draw valid conclusions about the effects of the acid on the samples tested. Record your observations at the end of 30 minutes and at 24-hour intervals for five days. Which materials were affected by the acid? Were any substances resistant to the acid?

3. For a long-term study, test the effects of the acidic solution prepared earlier by adding sulfur dioxide gas to distilled water; however, do not add the indicator. Another method to make a solution simulating acid rain involves adding carbon dioxide gas to distilled water. Can you think of a readily available source of carbon dioxide gas? After adding the acid to the various samples, cover each beaker with aluminum foil, and check for any changes in the appearance of the samples over a period of three to four weeks. Can you identify any local buildings, statues, headstones, or marble structures that have been corroded by acid precipitation?

4. To determine the effect of acid precipitation on living organisms, begin by preparing several hydrochloric acid solutions with different pH values. Beginning with 500 ml of 1N hydrochloric acid with a pH value of 0, dilute a 50 ml sample ten times with distilled water to make a pH 1 solution, ten times again for pH 2, ten times again for pH 3, and so on until you have hydrochloric acid solutions ranging from pH 0 to pH 6 (see Figure 11–4).

5. One aspect of life possibly affected by acid precipitation is seed germination. In order to germinate or sprout, seeds require moisture, oxygen, and the proper temperature. To determine the effect of acid rain on seed germination, evenly space several bean, corn, or pea seeds on absorbent paper cut to fit the bottom of a petri dish. You will need one petri dish with seeds for testing each hydrochloric acid solution.

6. After positioning the seeds, dampen the paper with the appropriate pH solution. Cover the seeds with another piece of absorbent paper cut to fit the top of the petri dish, and moisten with the same pH solution. Replace the cover and be sure to label each dish to keep track of the pH solution used.

7. Prepare another petri dish with seeds, but moisten the paper with distilled or rain water collected in a clean plastic container.

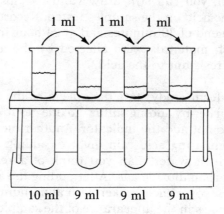

Figure 11–4. Adding 1 ml of 1N hydrochloric acid to 9 ml of distilled water diluting the solution ten times, producing a solution with a pH value of 2. Diluting again in the same manner results in a pH 3 solution, and so on.

8. On alternate days, remove the covers and examine the seeds in each dish. Record which solution is best for seed germination. Which pH solution has the greatest effect on seed germination? Once the seeds have sprouted, note which ones show the most growth as evidenced by the elongation of the stem and root. Remember to moisten the paper with the appropriate pH solution before replacing the cover. Continue your observations for two weeks.

Acids, Algae, and Animals

To assess the overall impact of acid precipitation on a wider scale, a small pond community can be established either at school or at home. This pond community can be used to explore the effects of acid precipitation on the interrelationships among organisms, including both animals and plants. Organisms living in a local pond, lake, or any habitat depend upon one another for nutrients to supply their energy needs. Changes in pH may have an immediate effect on one specific organism. However, others depending on the affected organism will eventually become victims of acid precipitation, leading to major imbalances among populations and possibly death for the entire community.

In the following experiment, you can set up several small aquaria to represent pond communities. By subjecting the organisms in each aquarium to a different pH solution, you can study the effects on an environment as it becomes increasingly acidic. In fact, you can explore the impact of acid shock by suddenly introducing a large amount of a more concentrated acid solution. This addition simulates the effect of the spring runoff from melting snow which can cause the acidity of a lake or pond to multiply by 100 times within a few days.

1. If small aquarium tanks are not available, three large glass containers are suitable for setting up the pond communities. Thoroughly wash some gravel, then put in enough to cover the bottom of the tanks. Mix in some plant fertilizer, which must contain nitrogen, phosphorus, and potassium. Also add a small amount of aluminum sulfate or magnesium sulfate to the gravel–fertilizer mixture.

2. Fill one tank with distilled water, the other two with any acid solution with a pH value of 4. Place the same combination of plants and animals in each of the tanks. The plants can include algae, duckweed, elodea, or any others obtained from a local pond or pet store. Be sure to include a variety of freshwater fish (especially live-bearers), snails, and other small aquatic animals.

3. Keep a daily record of the pH value for each aquarium. Since the fertilizer may make the distilled water slightly acidic, add some crushed limestone to neutralize this effect. Try to maintain the pH of the distilled water tank as close to 7 as possible throughout the experiment.

4. Allow the tanks to remain undisturbed for one week, recording any changes in the plant and animal populations. The acid solutions free the metal present in the sulfate compound. Subsequently, the aluminum or magnesium metal can be stored by plants and concentrated in any animal that feeds on them. To determine if the level of metals is increasing, compare the reproductive activity of the fish in each tank. Low reproductive rates and birth defects are characteristic of metal poisoning. The acid solutions may also cause nitrogen to be released from the fertilizer, leading to a proliferation of algae. With a explosive algae growth, oxygen is rapidly used, leaving little or none for other organisms living in the community. Is there any evi-

dence of an increase in the nitrogen level of the aquarium solutions?

5. At the end of one week, remove the water from one of the tanks with a pH 4 solution and replace it with a pH 2 solution, which is 100 times more acidic. Does this shock treatment have any immediate effect? Are any of the changes reversible if you replace the pH 2 solution with distilled water?

Test Your Local Rain

Having explored the chemistry of acid precipitation, why not check out the local rain for its pH value? Simply collect some rain in a plastic container that has been thoroughly washed, rinsed with 6N hydrochloric acid, and then rinsed several times with distilled water. After filtering the rain to remove any debris, check the acidity with universal pH paper. If a spectrophotometer similar to the one pictured in Figure 11–5 is available, you can analyze the rain sample for the levels of several substances known as *ions*. Unlike atoms, which are neutral, ions have either a positive or negative charge, depending on whether they have lost or gained electrons. With a spectrophotometer, you can test for magnesium, sodium, potassium, ammonium, sulfate, nitrate, and phosphate ions in your local rain. If positive tests are obtained, can you account for the sources of these ions? For a long-term project, compare the acidity of local rain at various times during the year. Is the rain more acidic during a particular season?

A Substitute for Charcoal

When it finally stops raining, your family heads outside again to enjoy another barbecue, only to discover that there's no propane gas in the tank to start a fire! A quick look in the garage reveals no charcoal, lighter fluid, or anything that might help—just some empty cans and a few chemicals left from previous chemistry experiments, including alcohol and some calcium acetate. Once again, you can save the day with a little practical knowledge of chemistry.

1. Prepare 25 ml of a saturated solution of calcium acetate.
2. Pour 50 ml of ethanol in a clean, dry metal container. While

Figure 11–5. A spectrophotometer can be useful in obtaining a detailed chemical analysis of water samples.

swirling the container, add 10 ml of the saturated calcium acetate solution. Be sure to add the calcium acetate solution all at once.

3. After the calcium acetate solution has been added, allow the container to remain undisturbed for several minutes. Record your observations.

4. Ignite the contents of the metal container and describe your observations. Can you think of a practical application for this material?

Uninvited Guests—Mosquitoes

With everything finally set for the family barbecue, another problem arises—mosquitoes! Without insect repellent, your family may finally give up any attempt to eat outdoors. Fortunately, a simple chemical recipe can provide a workable solution. Mix 10 gm of camphor and 10 gm of calcium chloride in 200 ml of ethanol.

Stir the ingredients to form a solution that can be applied to exposed areas of the skin to ward off mosquitoes and other annoying insects. With your knowledge of consumer chemistry, you can now predict the weather, prepare a heat source for cooking, and whip up an effective insect repellent. Any future family barbecues are guaranteed to be successful!

Topics for Further Investigation

Adding lime to lakes threatened by acid precipitation is one possible solution being considered. However, some scientists feel that the addition of lime to increase the pH would not be effective. Compare opposing viewpoints on this position. What is your recommendation?

The use of certain chemicals to control insects, notably DDT, has been banned in recent years. Investigate the chemical impact of these insecticides on both plants and animals. What chemicals are permitted to be used for insect control? Have they been shown to be safe for widespread application in areas subject to infestations by insects?

The trees in your backyard may have been destroyed by a relative newcomer to certain parts of the world, especially the North American continent—the gypsy moth. Ravenous foliage feeders, gypsy moths can devastate huge areas of forest land in a short time. Prepare a report on the history of this problem; include possible solutions.

Should one country be allowed to damage and possibly destroy another's ecology by emitting pollutants that lead to acid precipitation? Some countries, neighboring on more industrialized nations, are faced with this unsettled question. Should the offending country pay to clean up any damages?

Design experimental procedures to test the effects of acid precipitation on various life processes, for example, respiration, circulation, and embryological development. Include organisms exposed under natural conditions to acid precipitation.

Prepare a world map showing where acid precipitation is a concern. Indicate the severity of the problem by including diagrams to show the average pH values for particular regions.

*** Select a region where acid precipitation is a serious threat to the environment. Prepare an environmental impact study. Include sources of pollutants, long- and short-term effects, and specific management strategies and chemical treatment procedures to decrease the amount of emissions leading to acid deposition.

*** Using recombinant DNA technology, transform a bacterial species susceptible to acid precipitation into a strain that can live in solutions with low pH values.

12   *The Chemical Composition of Soil*

The soil from your lawn, garden, or backyard is a chemical factory. Containing almost every known element, soil is a mixture of minerals, salts, nutrients, gases dissolved in water, and organic compounds derived from animal wastes and decayed plants. More chemicals are added to the soil whenever fertilizers or weed killers are used. Moreover, the weathering action of the sun, rain, and wind causes the erosion of small rocks and stones, releasing additional minerals and salts into the soil. With a rich supply of numerous compounds, the soil is fertile ground for a variety of chemical reactions, all hidden beneath the surface of the earth. To begin an investigation of this chemical factory, take a shovel and fill up a small pail or bucket of some soil located near your house. If you live in an apartment, you can take soil samples from flower pots or from a patio herb garden.

Sandy or Silty Soil?

As you undoubtedly know, the vegetation growing in your soil depends upon water. But did you realize that most of the water falling on the ground from either rain or a garden hose drains away and does not get into the soil? Consequently, the small amount that does seep into the soil is vital for the growth and maturation of your plants, shrubs, fruits, and vegetables. However, the water must not only get into the ground but must also be retained by the soil for those times when it doesn't rain or whenever you forget to water the lawn or garden. The amount of water absorbed and retained by the soil depends upon its chemical composition.

For example, soil containing a high percentage of sand or gravel does not store much water. Instead, the water quickly passes down into deeper layers, carrying with it many of the nutrients needed for plant growth. On the other hand, soil with a high clay or silt content retains too much water. Such soils bind the water and dissolved nutrients, preventing their use by plants. With so much water, these soils become soggy and do not promote good growing conditions.

Plants thrive best in loam, a mixture of several substances, such as sand, clay, gravel, and silt, combined in the proper proportion for supporting plant growth. Many of the substances in loam are intermediate in size between the coarse particles in gravel and sand and the very fine grains in clay and silt. These intermediate-sized particles are colloids. As you may recall from the analysis of milk in Chapter 2, colloids are not small enough to dissolve or large enough to settle to the bottom of a solution. Instead, colloids remain evenly distributed throughout the solution. Much of the chemical action between the soil and plants occurs at the colloidal level.

The colloidal particles in loam function as a reservoir, retaining much of the water entering the ground. Whenever the amount of water seeping into the ground is insufficient to support plant growth, the colloidal particles in loam slowly release the water and dissolved nutrients they've held. Plants largely depend upon this water freed from the colloids in loam. Is there enough loam in your soil to provide proper growing conditions for plants? The following exercise can be done to investigate the type of soil used to grow vegetables in your garden or flowers on your patio.

How Much Loam in Your Soil?

1. Fill a 100-ml graduated cylinder to the 50-ml mark with your soil sample. Be sure that the soil is dry, not damp or moist from a recent rainstorm. Fill another cylinder to the same level with potting soil or loam obtained from a nursery or garden shop.
2. Add distilled water to the 100-ml mark. Place your hand on the top of the cylinder and shake vigorously for two minutes. Allow the cylinders to remain undisturbed overnight. As the soil settles, three distinct layers, consisting of coarse, medium, and fine particles, may form (see Figure 12–1).
3. The next day, record the height in milliliters of each layer for both soil samples. How does your local soil compare with the

Figure 12–1. After settling, what is the proportion of coarse, medium, and fine particles in your soil sample?

commercial sample? Is the height of the medium-sized particles the same for both samples?

4. Is the liquid layer above the soil clear or cloudy? To test for the presence of colloidal particles, shine a beam from a flashlight through this liquid layer. Describe your observations. If no colloids are present, the beam of light will pass straight through the solution. However, if colloids are in this liquid layer, they will scatter the light so that the beam is visible as it passes through the liquid. This diffusion of light by colloidal particles suspended in a solution is known as the *Tyndall effect.*

5. To check the capacity of your soil to absorb water, fill a 100-ml graduated cylinder to the 50-ml mark with soil thoroughly dried either in an oven or by the heat from the sun for several days. Fill another 100-ml graduated cylinder with tap water. Slowly add the water to the soil. Determine the volume of water that can be added before any rises above the soil. Repeat this

procedure with the potting soil or loam. Compare the volume of water absorbed by each soil sample.

6. Place 50 gm of dried soil in a clean beaker. To each sample, add 50 ml of tap water and mix with a glass stirring rod. Place the beakers near a window, and observe them every day until the soil samples are completely dry. Which soil dried the fastest? Which one retained water for the longest time?

How Sweet Is Your Soil?

The colloidal particles in soil affect not only its moisture content but also its pH level. In some areas, especially where oak and pine trees grow, the soil can become acidic as a result of the chemicals released from decaying leaves and needles left on the ground in the fall. The colloidal particles may neutralize some of this acid. However, if too much acid is deposited, you may have to add some lime, after removing the leaves, to "sweeten" the lawn and prepare the soil for the next growing season. Lime is made from calcium carbonate, a chemical often used to neutralize the acidity of soils. The calcium carbonate reacts with water trapped by the colloidal particles in the soil to produce a base, calcium hydroxide.

Many factors can affect the pH of soil, including the type of soil, the amount of rainfall, the brand of fertilizer used, the kinds of trees, shrubs, and crops grown, and the season of the year. The pH of soils in different locations can range from 4 to 11; in addition, fluctuations in the pH can be detected in soil taken from the same area at different times. These changes in the soil pH depend upon the nature of the chemical reactions occurring in the ground at any particular time of the day or season of the year. For example, as the moisture content increases after a rainstorm, the soil can become more acidic as the carbon dioxide gas reacts with water to produce carbonic acid. Conversely, the soil will become increasingly basic in drier weather as certain salts collect in the ground. By the way, if the soil becomes too basic, various substances, including sawdust, pine needles, and decomposed leaves, can be spread over the ground to lower the pH.

A soil neither too basic nor too acidic is required for a productive vegetable or herb garden. Most vegetables prefer a pH between 6 and 8 (see Figure 12–2). Within this range, many of the nutrients can be released from the soil particles, making them available to the vegetable plants. However, if you plan a flower

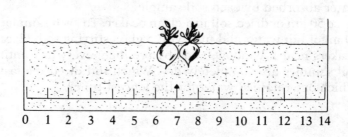

Figure 12–2. Vegetables grow best in a soil with a pH value between 6 and 8.

garden, be aware that some plants, including magnolias, holly bushes, azaleas, and gardenias, thrive best in an acidic environment. Obviously, a knowledgeable gardener may need one pH value for growing one kind of plant and a different soil pH for another type. But how can you be sure that the pH is correct for your purposes? The following experiment demonstrates how to determine the soil pH.

Does Your Soil Have the Proper pH?

1. Take several small samples (1 or 2 gm) from various areas of the garden or from different flowerpots and mix them in a clean, dry beaker.
2. Add enough distilled water to moisten the soil sample. Stir to get an even suspension of soil particles in water. Check the pH with universal indicator paper. Record the pH of the soil. If you are unsure whether the value is best suited for your purposes, check at a local nursery for the pH recommended for the plants you intend to grow. Determine whether the pH of your soil varies after a rainstorm or after any other climatic change.
3. If the pH of your soil sample is unsuitable, determine how much base or acid must be added to adjust the pH to the recommended value. Place 10 gm of the soil sample in a beaker. After adding 100 ml of distilled water, stir to get an even suspension. Check the pH with universal indicator paper.
4. Titrate the soil suspension by adding either 1N hydrochloric acid or 1N sodium hydroxide drop by drop until the pH is adjusted to the recommended value as determined by testing with a pH meter or universal indicator paper. How many ml of the acid or base were required to bring about the required change? Check at a local plant nursery to determine how much lime or

acid is recommended to adjust the pH value of your soil. Be sure to indicate the type of plant you intend to grow.

How Wet Is Your Soil?

The following investigation will enable you to determine the moisture content of a soil sample. Just as you may have done when checking the pH values, take soil samples exposed to varying climatic conditions to see how much the moisture content changes.

1. Place exactly 10 gm of soil in a clean, dry evaporating dish. Record the combined weight of the soil sample and dish.
2. Dry the soil in an oven at 60°C overnight. The next day, weigh the dish and soil. Subtract this value from the original one to determine the weight loss due to water evaporation. Determine the moisture content by placing your data in the following equation:

$$\frac{\text{weight lost (grams)}}{10 \text{ grams}} \times 100 = \text{percentage of moisture}$$

3. Compare various soil samples. Account for any differences observed in their moisture content.
4. Even after heating the soil in the oven, some water remains closely tied to the colloidal particles. These water molecules are held so tightly that they are not usable by plants. Can you think of a way to facilitate the release of these water molecules?

How Much Humus?

As the plants grow, older leaves and blossoms die, drop, and decay. The organic material contained in these plant parts is released into the soil, forming humus. As plants remove nutrients from the soil for their needs, the humus replenishes these materials and consequently serves as a major source of soil enrichment. Consequently, many substances are recycled between plants and the soil in which they are growing. Among these are starches, proteins, and lipids—organic compounds required for good growing conditions. Does your soil contain sufficient humus? The humus content of your soil can be determined by conducting the following exercise. You can then check at a local nursery to determine if the humus content is sufficient for your plants.

1. Record the weight of a clean, dry crucible.
2. Fill the crucible approximately halfway with soil and place it in an oven at 60°C overnight. What is the purpose of heating the soil sample at this temperature? The next day, describe the appearance of the soil.
3. Record the weight of the dried soil sample and crucible. Calculate the weight of the dried soil.
4. Place the crucible in a clay triangle supported by a ring stand (see Figure 12–3) and heat with a hot flame for 30 minutes. Stir the sample occasionally with a glass rod while heating. Be sure to maintain as hot a flame as possible.
5. Allow the crucible and contents to cool. Describe the appearance of the soil sample. Weigh the crucible and residue and determine how much weight has been lost by the soil sample. Assuming that this loss was due to the burning of the humus, calculate its percentage by placing your data in the following equation:

$$\frac{\text{weight lost (grams)}}{\text{weight of dried soil (grams)}} \times 100 = \text{percentage of humus}$$

6. What chemical substances would you expect to find in the ash remaining after the soil sample had been heated for 30 minutes?
7. Repeat this procedure using peat moss obtained from a local plant nursery or garden supply store.

Soil Is Never the Same

The chemical composition of humus is constantly changing, depending upon the kinds of leaves, flower parts, and branches that fall to the ground. In addition, the bacteria in the soil do not decompose large compounds into smaller molecules at the same rate all the time. As a result, chemicals present in humus are always in various stages of decay. Moreover, the chemical nature of humus varies to some extent depending upon the type of soil, weather conditions, and season of the year.

If the amount of certain chemicals in humus is adequate, then all the nutrients required for healthy plants will be provided. These chemicals include nitrogen for leaf and shoot growth, phosphorus for strong roots and stems, and potassium for preventing plant diseases. Other chemicals needed to a lesser extent are cal-

Figure 12–3. To determine the humus content of your soil, heat the sample in a crucible supported by a clay triangle.

cium, magnesium, sulfur, copper, zinc, iron, chlorine, and manganese. Of the more than 100 known elements, only 9 are required in any appreciable quantity and 5 in trace amounts for normal plant growth. These 14 elements, listed in the table below, play an essential role in plant nutrition. Relatively few elements are required for good growing conditions. How many of these are present in your soil?

ELEMENTS REQUIRED FOR PLANT GROWTH

Essential Elements

carbon
hydrogen
oxygen
calcium
iron
nitrogen
sulfur
potassium
phosphorus

Trace Elements

zinc
copper
boron
molybdenum
manganese

If the humus content of soil is inadequate, an additional source of these 14 elements must be provided. Either organic or inorganic fertilizers can be used to supply the missing nutrients. The addition of fertilizers is a direct method of supplying the soil with elements and minerals plants need but cannot obtain from the surrounding humus. Keep in mind, though, that applying fertilizer is only one requirement for healthy plants; the proper moisture content and pH are also important.

When additional soil nutrients are needed, many gardeners prefer organic fertilizers prepared from decayed matter instead of adding commercial products containing inorganic compounds. Under such circumstances, a person can claim that all plants have been grown using only natural conditions and ingredients. If you prefer such a method, then the necessary nutrients must be supplied from one of the following sources: compost, animal manure, fish-oil emulsions, or liquid seaweed preparations.

Compost is the most commonly used organic fertilizer. Formed from animal wastes, vegetable matter, lime, and water, compost is an effective way to recycle organic material left from both house and garden wastes. Not only does compost provide essential nutrients, but it also increases the ability of the soil to retain water and aids in the removal of salts that can damage plants. Making compost takes time but is a simple task, as described in the following procedure.

Make Your Own Compost

1. Using chicken wire, chain link or snow fencing, contain an area for developing a compost pile. Be sure to allow for air circulation and to prevent any scattering of material by wind or animals.
2. Add organic waste material to a depth of 1 foot to start the compost pile. Any organic material, including newspapers, paper bags, leaves, grass cuttings, kitchen wastes, and wood shavings, can be included. In fact, the material can be placed in paper bags or newspapers to eliminate the problem of attracting flying insects.
3. After spreading out the compost to a height of 1 foot, cover the surface with 3 inches of garden soil or dried manure to provide the bacteria needed for decomposing the organic compounds.
4. Keep the compost moist but not too wet. If the compost starts to have an unpleasant odor, add less water to the pile.
5. Repeat the layering of 1 foot of organic wastes and 3 inches of soil or manure (see Figure 12–4) until the fenced-off area is nearly full. After the area has been filled, allow it to remain undisturbed for six weeks. Add water when necessary. The chemical reactions brought about by the bacteria release energy in the form of heat. The temperature of a compost pile may reach 80°C because of the decomposition reactions.
6. After six weeks, turning is required to mix the various materials and promote contact between bacteria and materials not yet decomposed. To facilitate the mixing process, a second compost area is helpful. Remove the top layer from the first pile and place it on the bottom of the second one, continuing the process until all the layers have been inverted. The compost is ready when a small sample crumbles in your hand.
7. Relatively large amounts of compost are required in a garden— 6 inches spread over and then worked into the soil. Several layers should be applied, preferably just before planting time.

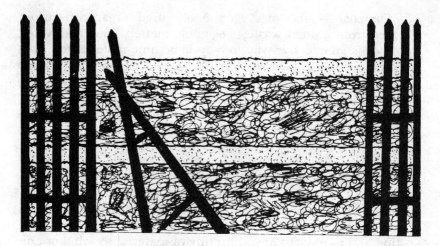

Figure 12–4. Alternating layers of organic wastes and soil (or manure) will produce good compost for garden use.

Your compost pile has turned wastes into a valuable source of plant nutrients!

Making Compost the Chemical Way

Many gardeners find preparing compost a time-consuming task that is not worth the effort because it does not provide as many plants nutrients as commercially prepared inorganic fertilizers. The best compost might contain 3 percent nitrogen, 1 percent phosphorus, and 2 percent potassium, whereas a commercial fertilizer can provide 5 percent, 10 percent, and 5 percent, respectively. Therefore, a much larger quantity of compost must be applied to a garden to supply the same levels of essential nutrients.

Another factor to be considered in comparing organic versus inorganic fertilizers is the moisture content of the soil. Compost material promotes water retention; consequently, any soil with a high moisture content may become too soggy to promote good growing conditions if decayed organic material is used as fertilizer. Certain ions, especially calcium and magnesium, are also retained by soils treated with organic compost. Consequently, these plant nutrients are not easily released by the colloids in the soil for use by the plants.

The proportion of nutrients in a fertilizer determines its suitability for your plants and soil. Required by law to label the percentage of the three principal nutrients (nitrogen, phosphorus, and potassium) on the package, the manufacturer indicates the percentage composition by a series of numbers. A fertilizer labeled 5–10–5 contains 5 percent nitrogen, 10 percent phosphorus, and 5 percent potassium (potash). Any fertilizer labeled 6–12–6 has the same proportion but a higher percentage of the three major plant nutrients.

Since the packaging of fertilizer is a factor in establishing its price, check for the percentage of each nutrient on the container. A 100-pound bag of 10–20–10 fertilizer may cost only one third more than a 100-pound bag of 5–10–5 fertilizer, yet each pound of the former supplies twice as many nutrients as one pound of the latter. Considering that only half as much volume is needed, your money buys more value with the 10–20–10 fertilizer!

 Comparing Fertilizers

The following computer program will compare any two commercial fertilizers to help you determine which is the best buy. After you enter the price, weight, and percentages of nitrogen, phosphorus, and potassium for each brand, the program will display an item analysis. Be sure to enter only the numbers—1.25 and not $1.25, 50 and not 50 pounds, 5 and not 5 percent, etc. For weight, the unit doesn't matter—pounds, grams, ounces—since only the number is used.

```
10 PRINT "WHAT IS YOUR FIRST NAME";:INPUT NA$
20 PRINT "NAME OF FIRST FERTILIZER";:INPUT FE$
30 PRINT
40 PRINT "PRICE";:INPUT PE
50 PRINT "WEIGHT";:INPUT WE
60 PRINT
70 PRINT "PERCENTAGE OF NITROGEN";:INPUT N
80 PRINT "PERCENTAGE OF PHOSPHORUS";:INPUT P
90 PRINT "PERCENTAGE OF POTASSIUM";:INPUT K
100 PRINT
110 PRINT "NAME OF THE OTHER FERTILIZER";:INPUT FR$
120 PRINT
130 PRINT "PRICE";:INPUT PR
140 PRINT "WEIGHT";:INPUT WT
150 PRINT
160 PRINT "PERCENTAGE OF NITROGEN";:INPUT NI
170 PRINT "PERCENTAGE OF PHOSPHORUS";:INPUT PH
180 PRINT "PERCENTAGE OF POTASSIUM";:INPUT KO
190 PRINT
200 PRINT "HERE IS THE COMPARISON,";NA$;"."
210 PRINT FE$;" COST PER UNIT WEIGHT: ";PE/WE
220 PRINT FR$;" COST PER UNIT WEIGHT: ";PR/WT
230 PRINT FE$;" HAS";N/NI*100;"%"
240 PRINT "AS MUCH NITROGEN."
250 PRINT FE$;" HAS";P/PH*100;"%"
260 PRINT "AS MUCH PHOSPHORUS."
270 PRINT FE$;" HAS"; K/KO*100;"%"
280 PRINT " AS MUCH POTASSIUM."
```

Fertilizers—Powders or Pellets?

No matter what the composition, fertilizers are available as powders, pellets, and concentrates. Be sure to dilute any concentrated brand to obtain the proper percentages of each nutrient. Any fertilizer left on foliage in a relatively high concentration can kill the plant, turning it brown and making it look burned. Pellets are more appropriate in drier weather because they won't be blown away as easily by wind. Since they don't stick to leaves, pellets are less likely to cause a burn when applied. Better yet, to avoid any of these potential problems, how about making your own chemical fertilizer?

Mix 3 gm of ammonium sulfate (for nitrogen), 6 gm of superphosphate (available from a plant nursery or garden supply store, needed for phosphorus), and 3 gm of potassium sulfate or potassium nitrate (for potassium) for every 4 liters of distilled water. Since they are dissolved in solution, the nutrients are readily available to your plants. The danger of burning your plants is greatly reduced because the fertilizer is diluted with water. You may want to add small amounts of insecticide, weed killer, and trace elements to your liquid fertilizer. An overdose of trace elements can be toxic to plants, so do not routinely include them when preparing your liquid fertilizer. In fact, it may be a good idea to take a soil sample to an agricultural station in your area to determine if any trace elements are required as supplements in the fertilizer before you include them in your preparation.

Ammonia in Fertilizer

The nitrogen in fertilizer is present as part of an ammonium compound. This ammonium can be converted into ammonia, which is easily detected by its pungent odor. You can test commercial fertilizers for the presence of ammonia by placing 1 gm of the sample in a test tube. Add 10 ml of 1N sodium hydroxide and carefully wave the vapors with your hand toward your nose. Can you smell ammonia? Since the vapors are toxic, do not inhale them. You can use this procedure to check other substances, including smelling salts, for the presence of ammonia or ammonium compounds.

Plant Growth Through Photosynthesis

With the right moisture content, proper pH, adequate amount of nutrients, and good drainage, conditions are perfect for plant growth. As your plants develop, they carry on a chemical process vital for all organisms, including both animals and other plants. Known as *photosynthesis*, this process requires carbon dioxide gas, water, sunlight, and chlorophyll, which is the pigment in plants responsible for their green color. Through a complicated series of chemical reactions, six carbon dioxide molecules are converted into one sugar molecule. The water serves as a source of oxygen gas. The sunlight provides the energy, which is trapped by the chlorophyll for use by the plants to keep photosynthesis operating.

To explore the chemical nature of photosynthesis, begin by placing 10 ml of distilled water in a test tube and add ten drops of bromothymol blue. Blow through a straw into the solution until a color change is apparent. What chemical compound from your breath produced this color change? With this information and a freshwater plant such as elodea, can you design an experiment to demonstrate that carbon dioxide is used by a green plant exposed to sunlight? By the way, bromothymol blue is not toxic to plants. In planning this experiment, be sure to include a control to prove that the plant was responsible for the disappearance of carbon dioxide gas from the solution.

With the same materials, design an experimental procedure, again with the proper control, to prove that light is necessary for the photosynthetic reaction. After your experiment has been running for at least 30 minutes, closely examine the plants placed in the solutions for the presence of small bubbles on their leaves. How can you account for the formation of these bubbles? Do these bubbles prove that photosynthesis is occurring? Why or why not?

A second way to determine if photosynthesis is taking place involves checking for another chemical product of the reaction— sugar. Since the sugar molecules are bonded to one another to form starch, test for starch using Lugol's iodine solution. First, boil a leaf from a geranium or any other broad-leafed plant in ethanol until all the chlorophyll has been extracted. After dipping the leaf in water to rinse off the ethanol, place it in the bottom half of a petri dish and cover with Lugol's iodine. The appearance of a blue-black color indicates the presence of starch. Test a variegated leaf in the same manner. Where does starch form in this type of leaf?

Plant Pigments

Boiling a leaf in ethanol extracts the chlorophyll pigment, as evidenced by the alcohol solution turning green. Actually, this extract is a mixture of different pigments, including various shades of greens, yellows, and purples. One technique often used to separate the various components in a mixture is *chromatography*. A sample of the pigment mixture is placed on specially treated paper, which is then placed in a solvent. As the solvent moves up the paper, each of the components in the mixture separates, because they do not move along at the same rate. Although chromatography was first used to isolate the various components in a mixture of pigments, this technique is now used to separate the individual parts in other substances. In the following exercise, try to see how many different colors can be obtained from a plant pigment extract.

1. Draw a pencil line 2 cm from the bottom of a piece of chromatography paper. Use forceps when working with the paper since the oils from your fingers can affect the results.
2. Dip a capillary tube into the pigment solution extracted in hot ethanol. Experiment with leaves and blossoms of various colors to see how many individual pigments can be isolated. Touch the capillary tube to the center of the pencil line and repeat the application until an intense color line overlays the pencil line (see Figure 12–5).
3. Pour the solvent, made from 5 parts acetone and 95 parts petroleum ether, into a large test tube. Do not insert the paper strip; however, make sure that the level of the solvent is below the

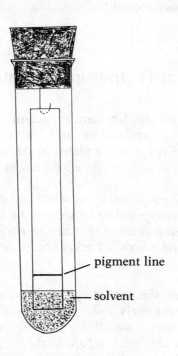

— pigment line

— solvent

Figure 12–5. To obtain a good separation, the plant pigment extract must not touch the solvent. Remove the chromatography paper when the solvent approaches the top.

pencil line drawn on the paper. When the paper is inserted into the test tube in the next step, the bottom edge of the chromatography strip must extend into the solvent.

4. With a paper clip, suspend the chromatography strip from a wooden cork so that it is properly positioned. Once you are sure that the bottom edge of the paper, but not the pigment line, will be immersed in the solvent, insert the strip. Be sure that the cork is tight.

5. Do not disturb the test tube. Observe the solvent move up the paper and record your observations. Remove the paper from the test tube when the solvent approaches the paper clip. After the chromatography strip is thoroughly dry, how many distinct pigment bands can you count? Explain why all these colors were not visible in the chlorophyll extract. Repeat this procedure with other materials, including india ink, food coloring dyes, and vegetable extracts. Also vary the proportion of the acetone–petroleum ether mixture to see if you can obtain a better separation of the individual pigments. After all, the end can be very colorful!

Topics for Further Investigation

In the nineteenth century, scientists discovered that plants could be grown without any soil. This technique, known as hydroponics, involves growing plants in salt solutions. Prepare a hydroponic solution and test its effectiveness for growing plants.

Sometimes plants grow when you don't want them to! Grass and weeds can sprout up between patio bricks and in rock gardens. Develop a salt solution that can be sprayed on these areas to kill the grass and weeds. Check some commercial products for their ingredients.

Although the basic steps of photosynthesis have been understood since the early 1900s, scientists are still working to unravel many of its mysteries. How is water split by the energy from light? What is the chemical nature of the compounds used in the light reaction? How can this process be accelerated to make plants grow faster? Plan an experimental procedure to answer one of these or any other unsolved question dealing with photosynthesis.

Obtain a soil sample that is unsuitable for plant growth, either because of its moisture or humus content, pH value, or lack of drainage and aeration. Plan a soil management program to improve the plant-growing capability of this soil.

Legumes are a family of plants that include peas, beans, and clover. The French, recognizing their importance, use the word *légume* for "vegetable." Farmers throughout the world grow these plants as winter crops. Why are legumes so valuable?

The science of agricultural technology is critical in those regions where malnutrition and hunger are serious problems. Contact an organization involved in this area and report on what methods are being used to improve the quality and quantity of food grown in these parts of the world.

Chromatography now involves silica gels and two-dimensional analyses. Prepare a chromatogram using these two features. Investigate gas chromatography and its application in chemistry.

*** Recombinant DNA technology is being applied to develop special strains of nitrogen-fixing bacteria that will grow on the roots of many different plants. Can you utilize this technique to develop a bacterial strain that is capable of changing free nitrogen into nitrates and is able to grow on nonleguminous plants?

*** With the help of plant genetics, develop a plant species that will grow under a wide variety of conditions. Consider plants that are important in world agriculture: wheat, soybeans, or corn.

Glossary

Acid—a compound that produces a pH value less than 7 when in solution

Amino Acid—an organic compound used as the building block of proteins

Atom—the smallest part of an element that can participate in a chemical reaction

Base—a compound that produces a pH value greater than 7 when in solution

Biodegradable—capable of being decomposed by the action of small organisms

Buffer—a chemical that helps to stabilize the pH

Calorie—the amount of energy required to raise the temperature of one gram of water by one degree Celsius

Calorimeter—an instrument used to measure the calorie content of a food sample

Carbohydrate—an organic compound characterized by the ratio of two hydrogen atoms to one oxygen atom

Chemical bond—the force uniting atoms to form a compound

Chromatography—a technique used to separate the various components in a mixture

Coagulation—the clumping of small particles to form a visible mass

Colloid—a mixture prepared by suspending one substance in a second substance

Compound—any substance that can be broken down into simpler substances by chemical means

Condensation—the formation of a liquid from a gas

Cracking—a process that breaks down larger hydrocarbons into smaller ones

Deionized water—water from which salts, minerals, and other dissolved materials have been removed

Density—the amount of mass, usually expressed as the weight, in a given volume

Desiccant—any substance capable of absorbing water

Disaccharide—a carbohydrate composed of two simple sugars

Distilled water—pure water from which all dissolved materials and contaminants have been removed

Electrolyte—any substance capable of conducting an electrical current

Electrons—negatively charged particles found in atoms

Element—a substance that cannot be broken down into simpler substances by ordinary chemical means; used as the building block for compounds

Emulsifier—a substance used to produce a colloid using two immiscible liquids

Emulsion—a colloid of two immiscible liquids in which small droplets of one liquid are suspended in a second liquid

Enzyme—a protein responsible for catalyzing or speeding up the rate of a chemical reaction

Ester—an organic compound characterized by a pleasant smell and formed by the reaction of an alcohol with an organic acid

Formula—a shorthand notation indicating the name and number of each element in a compound

Fractional distillation—the separation of individual components in a mixture based on differences in their boiling points

Hydrocarbon—an organic compound made entirely of hydrogen and carbon

Indicator—a substance that changes color in the presence of an acid or base

Inorganic—lacking carbon atoms as part of its chemical composition

Ion—a charged particle formed when an atom either gains or loses electrons

Lipid—an organic compound characterized by a high proportion of carbon and hydrogen atoms and a relatively low number of oxygen atoms

Mixture—a combination of substances that are not chemically combined; each substance retains its distinctive chemical properties

Molecule—the smallest part of any substance that retains all the chemical characteristics of that substance

Monosaccharide—a single sugar that cannot be broken down into a simpler carbohydrate

Mordant—a substance bonded to a fabric to facilitate the dyeing process

Normality—a measurement of the relative strength of an acid or base

Oil—(1) a lipid existing as a liquid at ordinary room temperature; (2) a product of fractional distillation of petroleum

Organic—a compound containing carbon, often in combination with hydrogen and oxygen atoms

Oxidation–Reduction—a chemical reaction involving the gain of electrons by one substance and the loss of electrons by another substance

pH—a measurement of the relative strength of an acid or base, ranging from 0 to 14

Photosynthesis—the chemical process by which green plants convert water into oxygen and carbon dioxide into sugar in the presence of light

Polymer—large compounds formed by bond formation of smaller units known as monomers

Polysaccharide—a carbohydrate composed of many simple sugars bonded to one another

Precipitate—solid material produced upon mixing two or more liquids

Protein—an organic compound composed of amino acids

Saponification—the reaction between a base and a fat that produces soap

Saturated compound—the presence of only single bonds in an organic compound

Solute—the substance that is dissolved in a solution

Solution—a uniform mixture of two or more substances

Solvent—the substance that does the dissolving to prepare a solution

Structural formula—the use of letters (representing elements) and lines (denoting chemical bonds) to illustrate the arrangement of the elements in a compound

Surface tension—the attraction between molecules within a liquid along the surface of the liquid

Titration—the process by which the concentration of an unknown solution is determined by the addition of a solution of known concentration until an end point is reached

Tyndall effect—the scattering of light by colloidal particles distributed in a solution

Unsaturated compound—the presence of one or more double bonds in an organic compound

Viscosity—the ability of a liquid to flow

Index